陳偉 編著

萬里機構・飲食天地出版社出版

在家做燒味

Homemade
Siu Mei
and Lou Mei

U0100088

前言

每次上烹飪課遇上新學員，話題都會涉及到菜式要怎樣做，他們會帶着疑惑的眼神，總是問我：「老師，我在家裏能否做到？」我回答：「當然可以。」他們投回的目光，還是半信半疑。到了下次再遇，他們大部份都很歡欣地說：「老師，我做過啦！真的成功了。」當然有小部份學員還沒有成功，他們會問：「問題究竟出在哪裏？做時要注意甚麼？」第三次重遇，他們十分雀躍地說：「我成功啦！」我知道，如果還不成功，他們再也沒興致來上課了。在這裏，我希望能跟大家分享一點心得，希望加強你們對烹飪的信心，可掌握烹調的竅門。

1. 當你想試做本書或講義上的新菜式時，必須專心看幾遍菜式的做法，特別要留心偉師傅的專業指導。

2. 食材、調味料或味料的份量必須精準，切勿一邊煮，一邊還在找尋調味料，這樣會錯漏百出，搞垮一切。

3. 每次完成一款菜式，無論好或壞，必定自我檢測，下回怎樣才能做得更好。倘若重複做四至五次，效果都不理想，試找出問題的關鍵，或提出疑問與導師研討，再做得不好，可能與經驗不足有關，多嘗試和多練習，總有一天會成功。

4. 如有空閒，多看一些和烹飪有關的資訊，拓展知識領域，提高烹調技巧。

有兩句話，必須緊記：「工多藝熟，精益求精」，這是絕對的！

期望能藉此書與你們一同探討烹調技藝，互動學習，誠心希望得到讀者們善意的批評和指導。

最後，本人也想為社會盡一點綿力，因此除掉正常開支，本書一切版權收益，將會轉送香港紅十字會，懇請諸君共襄善舉。謹此致謝！

陳偉

FOREWORD

I always came across the same questions from new cooking class students. They were always full of doubt when asking me, "Can I really do that at home?" "Yes, certainly!" However, they were still suspicious and I could tell from their eyes. When we met in the next lesson, they were all in joy. "I did it at home. It really worked." Well, few of them might not succeed. They will ask "How come? What's the problem? Anything I missed?" But when we met again in the third lesson, they were delighted. "I made it." I know they would have no interest to come again if they failed again. That's why I would like to share my experience with you. I hope that could increase your confidence in cooking and learn some cooking secrets.

1. When you try a recipe learnt from a book, please read a few times in advance. Please pay attention to "Points to remember" or "Tips".

2. Measure the ingredients, seasonings and condiments accurately and prepare everything in advance. Otherwise, you will make a lot of mistakes and spoil your whole cooking experience.

3. Please test and review after every trial no matter it was a success or failure. You can learn from experience. If you have practiced for 4-5 times but still not satisfied, try to find out the key point, or you can discuss with the chef. Practice make perfect. Keep trying!

4. In your leisure time, learn more about food and cooking. This will consolidate your knowledge and improve your cooking skill.

Remember "Practice makes perfect". This is absolutely true.

I hope this book can give us chance to learn and discuss together about cooking. I am sincerely looking forward to hearing your opinion.

As a contribution to our society, all copyright income of this book, after general expenses, will be donated to Hong Kong Red Cross. Let's join hands for charity. Thank you!

Chan Wai

前言 • FOREWORD　　2

特式燒滷
DELICACIES

畜肉類　MEAT

南乳叉燒　8
BBQ Pork with Red Sauce

焗爐烤乳豬　10
Oven Baked Suckling Pig

蜜汁叉燒腩　14
Honey Roasted BBQ Pork Belly

鹹香燒豬肘　16
Roasted Pork Knuckle

紅燒丁方肉　18
Braised Pork Cubes

蒜香脆皮骨　20
Crispy Roasted Spare Rib of Pork with Garlic

蒜香冰燒五層肉　22
Roasted Pork Belly with Garlic

泰式燒豬頸肉　24
Thai Style Roasted Pork Jowl

柱侯香芋火腩煲　26
Braised Taro and Roasted Pork with Chu Hou Sauce

魚香茄子順風煲　28
Sichuan Braised Eggplant and Pork Ear

蝦醬豆腐豬耳煲　30
Braised Tofu and Pig Ear in Shrimp Paste Sauce

口水雙膁　32
Braised Pork Knuckle and Beef Shank with Hot and Spicy Sauce

麻辣金肚絲　34
Shredded Ox Tripe in Hot and Spicy Sauce

泰汁金肚絲　36
Shredded Ox Tripe in Thai Sauce

五彩金肚絲　38
Shredded Ox Tripe with Vegetables

燒烤醬辣汁羊腩煲　40
Braised Lamb Brisket in Chili BBQ Sauce

禽肉類　POULTRY

香芒鵝片　42
Sliced Goose with Mango

鮑汁鵝掌翼煲　44
Braised Goose Web and Wing in Abalone Sauce

柱侯鵝掌翼煲　46
Goose Web and Wing in Chu Hau Sauce

南京鹽水鴨　48
Nanjing Simmered Duck in Salted Broth

堂剪妙齡鴨　50
Roasted Duckling

海蜇火鴨絲　54
Shredded Roasted Duck with Jelly Fish

沙爹燒鴨　56
Roasted Duck with Satay Sauce

圍村南乳鴨　58
Braised Duckling with Red Sauce

鮮果煙鴨胸　60
Smoked Duck Breast with Guava

湛江鹹香雞　62
Zhanjiang Savory Chicken

葱花麻油雞　64
Steamed Chicken in Sesame Oil

巧製奇津雞　66
Cajun Chicken

椒鹽雞軟骨　68
Sautéed Chicken Soft Bone in Spicy Salt

鮑汁鳳爪　70
Chicken Feet in Abalone Sauce

咖喱脆皮雞　72
Crisp Curry Chicken

金蒜蠔王燒雞　74
Roasted Chicken with Garlic in Oyster Sauce

香麻酥雞件　76
Spicy Chicken Nuggets

紅燒乳鴿　78
Roasted Pigeon

蒜香乳鴿煲　80
Braised Pigeon with Garlic

海產類　SEAFOOD

泰汁鮮果墨魚仔　82
Baby Squid in Thai Sauce with Fresh Fruit

沙薑墨魚仔　84
Baby Squid in Zedoary Sauce

鮑汁八爪魚　86
Octopus in Abalone Sauce

鹽香蜜汁燒鱔　88
Roasted Eel with Savory Honey Sauce

蜜汁燒帶子　90
Honey Roasted Scallop

涼菜小食
APPETIZER

葱花芥末海蜇頭　94
Jelly Fish with Mustard and Scallion

琥珀花生　96
Caramelized Peanuts

酸辣藕片　98
Lotus Root in Sour and Spicy Sauce

麻醬青瓜條　100
Cucumber Strips with Sesame Paste

七味涼瓜　102
Bitter Melon in Spicy Sauce

香麻萵筍絲　104
Celtuce Shreds in Sesame Sauce

欖菜四季豆　106
String Beans with Preserved Cabbage

金磚脆豆腐　108
Pan Fried Tofu

葱花木魚凍豆腐　110
Tofu with Scallion and Bonito Flakes

冰梅皮蛋酸薑　112
Preserved Egg with Pickled Ginger in Plum Sauce

常用香料和醬料
SPICES AND SAUCES

常用的香料　116
Spices

味料和上皮料的配製　121
Seasonings and Brine

附錄 | 燒味行話　126
Appendix | Jargons

【特式燒滷】

DELICACIES

南乳叉燒

BBQ Pork with Red Sauce

份量：**8-10位用** ■製作時間：**90分鐘** ■工具：**鋼針4枝**

■ Serves: **8-10**

■ Preparation and Cooking Time: **90 minutes**

■ Utensil: **4 stainless steel skewers**

材料：
枚肉2斤（1.2千克）
蜜汁8兩（300克）

味料：
乾葱茸3粒
蒜茸6粒
砂糖5湯匙
精鹽1湯匙
雞粉1湯匙
麻醬1湯匙
大南乳1件，壓爛成茸
美極鮮露2湯匙
雞蛋2隻
五香粉⅓茶匙
玫瑰露酒1湯匙

蜜汁：
麥芽糖12兩（450克）
清水4湯匙
冰糖2兩（75克）
薑2片
精鹽1湯匙

Ingredients:
1.2kg tenderloin pork
300g honey sauce

Marinade:
3 cloves shallots, chopped
6 cloves garlic, minced
5 tbsp granulated sugar
1 tbsp salt
1 tbsp chicken powder
1 tbsp sesame paste
1 pcs large fermented red bean-curd, crashed
2 tbsp Maggi seasoning
2 eggs
⅓ tsp five spices powder
1 tbsp rose wine

Honey Sauce:
450g maltose
4 tbsp water
75g rock sugar
2 ginger slices
1 tbsp salt

◥ TIPS 偉師傅的專業指導

1. 因為枚肉含水量多，所以一定要瀝乾，否則難以吸收味料，使叉燒味道不足。

2. 焗爐分為煤氣爐和電爐，煤氣的最高溫是250℃，電爐則有280℃。

1. The BBQ Pork will not taste good if the meat does not drain to dry enough. It is because tenderloin is very juicy and it can't fully absorb the flavor of the marinade.

2. There are 2 types of oven: gas oven and electrical oven. The highest temperature for gas oven is 250℃. The highest temperature for electrical oven is 280℃.

做法：

1. 把蜜汁材料放進碗裏，隔水煮融，拌勻後煮滾，備用。

2. 將枚肉切成長條形，洗淨，瀝乾。

3. 把味料混合拌勻，放入枚肉，抹勻，醃30分鐘，每隔10分鐘翻動1次。

4. 用鋼針把枚肉串起，放進已預熱的焗爐裏，用280℃高溫烤15分鐘，取出，翻轉至另一面，放回焗爐再烤10分鐘。

5. 取出稍微放涼，抹上蜜汁，放回焗爐以180℃中火烤15分鐘。取出，並將四周焦邊剪去，再抹上蜜汁便成。

Method:

1. Put all the honey sauce ingredients in a bowl, sit over a hot water bath to melt. Bring to the boil. Set aside.

2. Cut the pork into long stripes. Wash. Drain. Set aside.

3. Marinate the pork with marinade for 30 minutes. Turn the pork every 10 minutes.

4. Thread the meat onto stainless steel skewers. Roast them in a preheated oven for 15 minutes at 280℃. Remove from the oven. Turn the meat over and roast again for 10 minutes.

5. Leave to cool for a while. Blast with honey sauce. Roast in the oven for another 15 minutes at 180℃. Trim the burnt and glazed with some honey sauce. Serve.

焗爐烤乳豬

Oven Baked Suckling Pig

份量：**12-14位用** ■製作時間：**11小時**
■ Serves: **12-14**
■ Preparation and Cooking Time: **11 hours**

材料：
小乳豬1隻約6斤（3.6千克），俗稱「席豬」

味料：
精鹽4湯匙
雞粉1湯匙
五香粉 ⅟₁₀ 湯匙
沙薑粉 ⅟₁₀ 湯匙

上皮料：
白醋2兩（75克）
浙醋1茶匙
麥芽糖 ½ 茶匙
玫瑰露酒 ⅟₁₀ 茶匙

蘸醬：
燒烤醬3湯匙
海鮮醬3湯匙
麻油 ½ 湯匙

Ingredients:
1 suckling pig (3.6kg)

Seasonings:
4 tbsp salt
1 tbsp chicken powder
⅟₁₀ tbsp five spices powder
⅟₁₀ tbsp zedoary powder

Brush-on Glaze:
75g white vinegar
1 tsp red vinegar
½ tsp maltose
⅟₁₀ tsp rose wine

Dipping:
3 tbsp BBQ sauce
3 tbsp Hoi Sin sauce
½ tbsp sesame oil

焗爐烤乳豬
Oven Baked Suckling Pig

做法：

1. 把蘸醬調和，一同煮滾，待用。
2. 把乳豬放入已加竹笪的鍋子，用大滾水焯10分鐘，取出泡冷水，瀝乾。
3. 把味料拌勻，放進豬肉裏抹勻。
4. 移去豬的鐵線、鋁桿、豬叉，分成四份，放進已預熱焗爐裏，以約200℃中火烤10分鐘，再用約280℃大火烤15分鐘，取出，稍涼後切塊，食用時跟蘸醬汁便成。

Method:

1. Blast the pig evenly with brush-on glaze. Set aside. Mix the dipping and bring to boil. Set aside.
2. In a big wok, put a bamboo mat. Blanch the suckling pig in boiling water for 10 minutes. Drain.
3. Blast the pig with seasonings.
4. Remove all the wires, forks and rods. Chop the pig into 4 pieces. Roast the pig in the preheated oven at 200℃ for 10 minutes over medium heat. Roast for 15 minutes at 280℃ over high heat. Remove from oven. Leave to cool for a while. Chop the pig into small pieces. Serve with dipping.

◥ TIPS 偉師傅的專業指導

1. 市面有多間凍肉批發商，供應已切好的乳豬。
2. 烤乳豬工具──不銹鋼小豬叉1隻；（A）上豬鋁桿1套：長24吋 × 寬2吋 × 厚½吋，共1條。（B）2條長6吋 × 寬2吋 × 厚½吋鋁桿，1條10吋長鐵線；1條6吋鋼針。
3. 烤乳豬時，爐溫如不夠高，須在烤熟後，用新鮮油炸脆乳豬皮。
4. 家用式焗爐的爐火較小，所以上皮料糖份要比手動轉架燒乳豬的上皮料較高，因而較容易起珠皮和易燒至快起色。

1. Suckling pigs can be bought from frozen food supplies.
2. Roast suckling pig tools—1 pig fork,1 set of rod for positioning: (A) 1 rod of 24 inches long x 2 inches wide x ½ inch thick, (B) 2 rods each of 6 inches long x 2 inches wide x ½ inch thick, 1 wire of 10 inches long, 1 skewer of 6 inches long.
3. If the oven temperature is not high enough, you need to deep fry the pig to made the skin crispy after roasting.
4. As household used oven is not as powerful as industrial oven, we need thicker. It is easier to get burnt.

串燒叉法

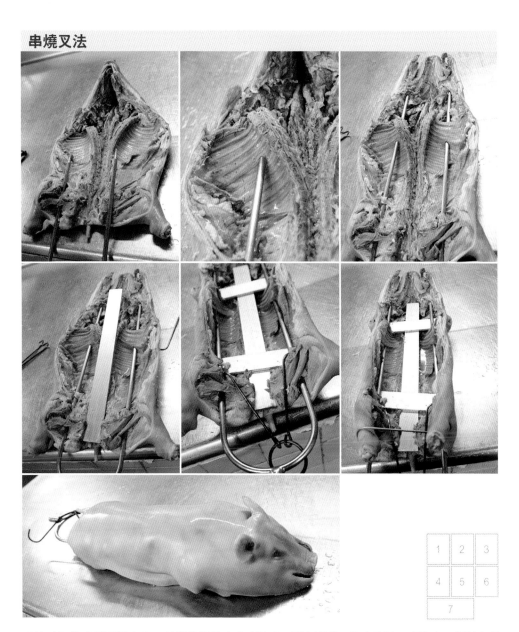

1	2	3
4	5	6
7		

用燒豬叉從豬腳兩側穿入，並從後腿骨穿出（圖1），到中間第4條肋骨插入（圖2），穿到豬頭兩邊的牙床位穿出（圖3），中間直架起上豬鋁桿A使豬身平直（圖4），近豬肘位和後脾位橫架鋁桿B使豬身定形（圖5），用鐵線把豬腳紮緊（圖6）。串好後洗淨，瀝乾，在豬皮均勻地抹上皮料，然後風乾10小時（圖7）。

Thread in at the side of the pork leg and thread out from the back leg bone with a big fork (Fig.1). Then thread in at the 4th spare rib (Fig.2) and thread out from the jaws (Fig.3). Put rod "A" in the middle of the body in order to keep the body flattened (Fig.4). Put rod "B" at the pig's knuckles and hips to position the shape (Fig.5). Tie the buttom with wire (Fig.6). Wash the pig thoroughly. Drain. Blast the body with brush-on glaze. Air-dry for 10 hours (Fig.7).

蜜汁叉燒腩

Honey Roasted BBQ
Pork Belly

份量：**8-10位用** ■製作時間：**80分鐘**

■ Serves: **8-10**

■ Preparation and Cooking Time: **80 minutes**

材料：

新鮮去骨五花腩肉4條，共2斤
（1.2千克）
蜜汁1碗（參閱第122頁）

味料：

砂糖6湯匙
精鹽1½湯匙
雞粉1湯匙
五香粉¹⁄₁₀湯匙
沙薑粉¹⁄₁₀湯匙
燒烤醬2湯匙（參閱第121頁）
生抽2湯匙
雞蛋2隻
玫瑰露酒2湯匙
蒜茸6粒
乾葱茸4粒

Ingredients:

4 pcs pork belly, fresh and boned (1.2kg)
1 bowl honey sauce (refer to p.122)

Marinade:
6 tbsp granulated sugar
1½ tbsp salt
1 tbsp chicken powder
¹⁄₁₀ tbsp five spices powder
¹⁄₁₀ tbsp zedoary powder
2 tbsp BBQ sauce (refer to p.121)
2 tbsp light soy sauce
2 eggs
2 tbsp rose wine
6 cloves garlic, minced
4 cloves shallot, minced

TIPS 偉師傅的專業指導

醃腩肉的時間切勿多於1小時，否則豬肉會被味料密封，肉質纖維變得太粗糙太韌，不夠嫩滑。

Pork belly should not marinated for more than an hour. Otherwise, the brine will seal the meat. The meat tissue will become very rough and loss the tenderness.

做法：

1. 把五花腩洗淨，瀝乾，放入味料拌勻，醃40分鐘，每隔10分鐘翻動1次，方便肉質吸入調味。
2. 用鋼針把腩肉串起，放上鐵架上。
3. 把焗爐預熱，放入五花腩以250℃-280℃高溫烘烤10分鐘，取出翻過來，再烘烤10分鐘。
4. 把火調至約200℃-230℃，烤烘10分鐘，取出稍微放涼，抹上蜜汁，再回焗爐以約200℃烘烤5分鐘，取出翻過來，繼續烘烤5分鐘，取出，稍涼後再抹上蜜汁便成。

Method:

1. Wash and dry the pork. Marinate it for 40 minutes. Turn the meat every 10 minutes in order to absorb the marinade.
2. Thread the meat with steel skewers.
3. Preheat the oven. Roast the meat in the oven over high heat at 250℃-280℃ for 10 minutes. Turn over and roast for another 10 minutes.
4. Roast the meat at 200℃-230℃ for 10 minutes. Leave to cool for a while. Glaze the meat with some honey sauce. Roast again for 5 minutes at 200℃. Turn over and roast for 5 minutes. Glaze with honey sauce again. Serve.

鹹香燒豬肘

Roasted **Pork Knuckle**

份量：**8-10位用** ■製作時間：**6小時**
■ Serves: **8-10**
■ Preparation and Cooking Time: **6 hours**

材料：

德國豬手2隻，約2斤（1.2千克）

味料：

八角2粒
草果2粒
香葉10片
丁香3粒
沙薑6粒
清水10斤（6千克）

上皮料：

白醋2湯匙
浙醋¼茶匙
麥芽糖⅛茶匙
紹酒¼茶匙

蘸醬：

泰國雞醬5湯匙

Ingredients:
2 pork knuckles (1.2kg)

Marinade:
2 star anise
2 brown cardamom pods
10 bay leaves
3 cloves
6 zedoary
6 kg water

Brush-On Glaze:
2 tbsp white vinegar
¼ tsp red vinegar
⅛ tsp maltose
¼ tsp Shaoxing wine

Dipping:
5 tbsp Thai sweet chili sauce

做法：

1. 把豬皮上皮醬料調和，待糖完全溶解。

2. 豬手解凍後，先汆水1次，泡冷水，放入味料裏煮滾，轉文火熬50分鐘，熄火，浸泡50分鐘，取出，用刀起出大骨後，抹去豬皮上的油漬，並均勻地抹上上皮料，掛起，風乾3小時。

3. 把油煮到大約6成熱，轉用文火，慢慢地澆在豬皮上；將豬手放在罩籬上，放入油鍋裏。

4. 待油溫升至9成熱時，察看全隻豬手是否呈金黃色，若豬皮表面出現芝麻皮，變得鬆脆，便成。食用時蘸泰國雞醬。

Method:

1. Mix the ingredients of the brush-on glaze. Stir well until the maltose is dissolved.

2. Defrost the knuckle. Scald the knuckles. Add into the marinade and bring to boil. Simmer for 50 minutes. Turn off the fire and let the knuckle soaked for 50 minutes. Take out the knuckle. Remove the bone with knife. Wipe away any oil on the knuckle. Blast with the brush-on glaze. Hang up and air-dry for 3 hours.

3. Put the knuckle on a Chinese ladle and then put it into the oil. When the oil is about medium heat, lower the fire. Pour oil over the knuckle slowly and constantly.

4. When the oil is going to burnt, alert the color of the knuckle if it is golden brown or not. If there are charcoals occurred on the skin and it looks crunchy, it means the food is ready. Serve with the Thai sweet chili sauce.

紅燒丁方肉

Braised **Pork Cubes**

份量：**20位用** ■製作時間：**5小時**
■ Serves: 20
■ Preparation and Cooking Time: **5 hours**

材料：
新鮮挑骨連皮豬腩肉5斤（3千克）
芫荽葉2棵
紅椒絲1隻

味料A：
蒜茸10粒
乾葱茸10粒
陳皮茸1塊
燒烤醬2湯匙
豆瓣醬2湯匙
日本燒肉汁6湯匙
草果4粒
紅麴米1湯匙，拍碎
胡椒粉¼茶匙

味料B：
蠔油2湯匙
美極鮮露2湯匙
喼汁2湯匙
冰糖5兩（188克）
精鹽1湯匙
雞粉2湯匙
紹酒2湯匙
玫瑰露酒2湯匙
清水6斤（3.6千克）

Ingredients:
3kg pork belly with skin, fresh and boned
2 stks parsley
1 red chili pepper, julienne

Seasonings A:
10 cloves garlic, minced
10 cloves shallot, minced
1 pcs dried tangerine peel, fine chopped
2 tbsp BBQ sauce
2 tbsp chili bean paste
6 tbsp Japanese BBQ Sauce (Yakiniku)
4 brown cardamom pods
1 tbsp red yeast rice, crushed
¼ tsp pepper

Seasonings B:
2 tbsp oyster sauce
2 tbsp Maggi seasoning
2 tbsp Worcestershire sauce
188g rock sugar
1 tbsp salt
2 tbsp chicken powder
2 tbsp Shaoxing wine
2 tbsp rose wine
3.6kg water

TIPS 偉師傅的專業指導

要預先炆好丁方肉，然後放兩小時，才翻熱來吃，便能讓味料慢慢滲入肉裏，達到香醇、軟滑的效果。

In order to bring out the flavor and ensure the meat is tender, we need to let the stewed meat put aside for 2 hours. Reheat the meat to eat.

做法：
1. 把腩肉切成3吋×3吋的方塊，洗淨瀝乾，放入約220℃滾油炸至底部和四周微焦，取出。
2. 在大煲底墊2張竹笪，平排腩肉，切勿重疊。
3. 燒熱鍋，炒香味料A，加入味料B煮滾，放進腩肉煮滾，轉慢火炆2小時，熄火，加蓋放2小時。
4. 食用前先翻熱，約煮30分鐘，把汁液煮稠，淋在肉上，撒上芫荽葉和紅椒絲便成。

Method:

1. Cut the pork into 3 inches x 3 inches cubes. Wash and dry. Deep fry in oil at 220°C until become light brown. Set aside.
2. Place 2 bamboo mats on a big wok. Lay the meat nicely. No stacking.
3. Heat the wok, sauté seasonings A till fragrant. Add seasonings B. Bring to a boil. Add the meat. Simmer for 2 hours at low heat. Turn off the fire and covered the wok with a lid. Leave the pan for 2 hours.
4. Preheat before serving, cook the meat for 30 minutes. Thicken the sauce and pour it on the meat, garnish with parsley and shredded chilli on the top if you need.

蒜香脆皮骨

Crispy Roasted Spare Rib of
Pork with Garlic

份量：**10-12位用** ■製作時間：**5.5小時**
■ Serves: **10-12**
■ Preparation and Cooking Time: **5.5 hours**

材料：
腩排2斤（1.2千克）
鹼水1湯匙

味料A：
蒜肉4兩（150克）
清水4兩（150克）

味料B：
砂糖 1/10 湯匙
精鹽 2/5 湯匙
雞粉 2/5 湯匙
鬆肉粉（安哥夫）1/10 湯匙
生粉 1/2 湯匙
澄麵 1/2 湯匙
蛋白1隻

Ingredients:
1.2kg pork ribs
1 tbsp lye water

Seasonings A:
150g garlic
150g water

Seasonings B:
1/10 tbsp granulated sugar
2/5 tbsp salt
2/5 tbsp chicken powder
1/10 tbsp tenderizer
1/2 tbsp cornstarch
1/2 tbsp wheat flour
1 egg white

◣ TIPS 偉師傅的專業指導

腩排下油鍋時，油溫不能太低也不能太高。油溫太低會將粉料沖散；油溫太高則很快會把腩排表面炸焦。

The oil should be medium hot in the beginning. If the oil temperature is too low, the ribs could not be nicely coated. However, if the oil temperature is too high, the ribs will be burnt outside but undercook inside.

做法：
1. 把腩排切成12件長條形，再與鹼水拌勻，醃5分鐘，取出，放在水喉下沖洗1小時備用。
2. 用攪拌機把味料A打成茸汁，將腩排放到味料A汁裏拌勻，醃1小時，濾去汁料，加入味料B拌勻，醃4小時，每隔1小時翻動1次。
3. 把油煮至6成熱，放進腩排，先用慢火浸熟，然後轉用大火炸硬，表面呈現金黃色，取出。

Method:
1. Chop the pork ribs into 12 pieces. Marinate with lye water for 5 minutes. Wash under running water for an hour. Set aside.
2. Blend seasonings A with a blender. Marinate the ribs in seasonings A for 1 hour. Drain the marinade. Marinate again with seasonings B for 4 hours. Stir the mixtures every hour.
3. Heat the oil until medium hot. Put in the ribs. Simmer the ribs till cooked. Turn to high heat and fry the ribs until golden brown.

蒜香冰燒五層肉

Roasted Pork Belly
with Garlic

份量：**20位用** ■製作時間：**3.5小時**
■ Serves: **20**
■ Preparation and Cooking Time: **3.5 hours**

材料：
去骨帶皮腩肉5斤（3千克）

上皮料：
鬆肉粉 ½ 茶匙
精鹽 3 湯匙

味料：
熟蒜茸 1 湯匙
生蒜茸 1 湯匙
砂糖 2 湯匙
精鹽 4 湯匙
雞粉 1 湯匙
五香粉 ⅛ 湯匙

Ingredients:
3kg pork belly with skin, boned

Brush-On Glaze:
½ tsp tenderizer
3 tbsp salt

Marinade:
1 tbsp garlic, cooked and minced
1 tbsp garlic, minced
2 tbsp granulated sugar
4 tbsp salt
1 tbsp chicken powder
⅛ tbsp five spices powder

◥ TIPS 偉師傅的專業指導

鬆針時力度要控制得宜，厚皮的部份不妨大力一點；薄皮部份則要輕柔。

Be careful when pricking the skin. Prick harder where the area is thick and gentle at thin layer.

做法：

1. 把味料拌勻，備用。

2. 把腩肉洗淨瀝乾，均勻地在豬皮上抹上皮料，翻過來，在骨的隙縫上割數刀，抹上味料，醃2小時。

3. 把腩肉用鋼針串成井字形，放入已預熱200℃的焗爐裏烘烤10分鐘，待豬皮變成媽紅色時取出，用針在豬皮上刺插，俗稱鬆針，掛在風口吹涼。

4. 焗爐預熱後，放入腩肉，以中火200℃燒8分鐘，取出再次刺插，放回焗爐，轉280℃大火燒約15-20分鐘，待腩肉的皮全部燒焦，表面滲出油時，蓋上一張錫紙，調至120℃火力，再烤25-30分鐘，取出。

5. 將燒焦的豬皮刮走便成。

Method:

1. Mix the seasonings. Set aside.

2. Wash the pork belly. Drain and dry. Blast the skin with brush-on glaze. Turn over the meat and have a few cuts between bones. Blast with seasonings and marinate for 2 hours.

3. Thread the meat with steel skewers into a square field shape. Roast the meat in preheated oven at 200°C for 10 minutes. When the pork skin looks red, remove from oven. Prick on the skin with steel needle and leave to cool.

4. Preheat the oven. Roast the meat in the oven at 200°C for 8 minutes. Remove the meat from oven. Prick on the skin and roast for 15- 20 minutes at 280°C. When oil sweats on the burnt skin, cover the meat with an aluminum sheet. Roast again for 25-30 minutes at 120°C. Remove from oven.

5. Before serving, scrap away any burnt on the skin.

泰式燒豬頸肉

Thai Style Roasted
Pork Jowl

份量：**1.2千克** ■製作時間：**1.5小時**

■ Serves: **1.2 kg**

■ Preparation and Cooking Time: **1.5 hours**

材料：

新鮮豬頸肉2斤（1.2千克）
紅椒1隻，切絲，墊底用
西生菜或椰菜1個，切絲，墊底用

味料：

砂糖3湯匙
精鹽1湯匙
生抽1½湯匙
玫瑰露酒1湯匙
雞粉1湯匙
魚露1湯匙
胡椒粉⅒湯匙
指天椒茸2隻
蒜茸6粒
乾葱茸3粒
柱侯醬或自製燒烤醬1½湯匙

蘸汁：

蒜茸2粒
乾葱茸2粒
指天椒茸3隻
青檸汁1個
白醋3湯
浙醋1湯匙
生抽½湯匙
魚露1湯匙
麻油½湯匙
清水6湯匙
泰國雞醬3湯匙

Ingredients:
1.2kg Pork Jowl
1 red chili pepper, julienne
1 lettuce or cabbage, shredded

Marinade:
3 tbsp granulated sugar
1 tbsp salt
1½ tbsp light soy sauce
1 tbsp rose wine
1 tbsp chicken powder
1 tbsp fish sauce
⅒ tbsp pepper
2 red cluster peppers, chopped
6 cloves garlic, minced
3 cloves shallot, minced
1½ tbsp Chu Hau sauce or home-made BBQ sauce

Dipping:
2 cloves garlic, minced
2 cloves shallot, minced
3 red cluster peppers, chopped
1 Kaffir lime , use the juice
3 tbsp white vinegar
1 tbsp red vinegar
½ tbsp light soy sauce
1 tbsp fish sauce
½ tbsp sesame oil
6 tbsp water
3 tbsp Thai sweet chili sauce

◤ TIPS 偉師傅的專業指導

1. 自製燒烤醬，已在附錄裏記載，請參考第121頁。

2. 在一般超市或街市雜貨鋪可買到柱侯醬。

1. The recipe of home-made BBQ honey sauce can be found at appendix on p.121.

2. We can buy Chu Hau sauce in supermarkets or grocery stores.

做法：

1. 混合蘸食料，煮滾備用。拌勻味料，備用。

2. 豬頸肉洗淨，瀝乾，放入味料拌勻，醃40分鐘，每隔10分鐘翻動1次。

3. 焗爐預熱，豬頸肉放鐵架上入以高溫火280℃烤15分鐘，取出，翻過來，回焗爐燒15分鐘，取出，剪去肉旁的燒焦部份。

4. 在盤上墊椰菜絲和紅椒絲，放上豬頸肉，蘸汁亨用。

Method:

1. Mix the dipping. Bring to boil. Set aside. Mix the seasonings. Set aside.

2. Wash the meat. Drain and dry. Mix with seasonings and marinate for 40 minutes. Turn the meat every 10 minutes.

3. Preheat the oven. Roast the meat for 15 minutes at 280℃ over high heat. Turn over the meat and roast again for 15 minutes. Trim the burnt.

4. Place shredded vegetables and pepper on a plate. Put the meat on top. Serve with the dipping.

柱侯香芋火腩煲

Braised Taro and Roasted Pork with **Chu Hau Sauce**

份量：**4-6位用** ■製作時間：**20分鐘**

■工具：**7吋瓦煲1個（預熱）**

■ Serves: **4-6**

■ Preparation and Cooking Time: **20 minutes**

■ Utensil: **1 clay-pot, 7-inch in diameter (pre-heated)**

材料：
燒腩仔6兩（225克）
荔芋8兩（300克）
蔥絲1條
紅椒絲¼隻

味料：
清湯10兩（375克）
雞粉½湯匙
砂糖½湯匙
麻油¼茶匙
蠔油¼湯匙
柱侯醬1湯匙
老抽¼湯匙
生薑2片
八角1粒
蒜肉1粒
洋蔥¼個

芡汁：
生粉½茶匙＋清水2湯匙（調勻）

Ingredients:
225g roast pork belly
300g taro
1 stk scallion, shredded
¼ red chili pepper, shredded

Marinade :
375g broth
½ tbsp chicken powder
½ tbsp granulated sugar
¼ tsp sesame oil
¼ tbsp oyster sauce
1 tbsp Chu Hau sauce
¼ tbsp dark soy sauce
2 slices ginger
1 star anise
1 clove garlic
¼ onion

Thickening:
½ tsp cornstarch + 2 tbsp water

◤ TIPS 偉師傅的專業指導

1. 如要效果更好，可多加4兩（150克）去皮荔芋，洗淨，隔水蒸20分鐘，用刀壓成茸，代替芡汁，香芋味道就更濃郁了。

2. 瓦煲要先預熱，這樣放進物料時便會冒煙、發出聲響，讓食物無論在色澤和味道上都會更加誘人。

1. Instead of adding the thickening, taro purée can rich the flavor too. To make that purée, peel 150g taro, wash and steam for 20 minutes. Mash the taro by a knife.

2. The clay-pot must be preheated. When the food is transferred to the pot, the sizzling sound will make the dish more presentable.

做法：

1. 荔芋去皮洗淨，切成長1吋 × 闊½吋 × 厚¼吋的長塊形，放滾油炸熟，盛起備用。

2. 燒腩仔切成長1吋 × 厚½吋的薄片，用慢火把兩邊煎香。

3. 在鍋裏下油，炒香生薑、蒜肉、洋蔥、八角和柱侯醬，加入其他味料以慢火煮2分鐘，濾去味料。並預熱瓦煲。

4. 放進荔芋和火腩同煮2分鐘，待汁液煮稠，倒下芡汁，放進瓦煲，撒上蔥絲和紅椒絲便成。

Method:

1. Peel the taro. Wash. Cut into slices of about 1 inch long x ½ inch wide x ¼ inch thick. Deep fry. Drain and set aside.

2. Slice the roasted pork into 1 inch long x ½ inch thick. Fry them in low heat.

3. Oil the wok. Sauté the ginger, mince garlic, onion, star anise and Chu Hau sauce till fragrant. Add other seasonings and cook over low heat for 2 minutes. Strain the mixtures. Preheat the clay-pot.

4. Add the taro and meat. Cook for 2 minutes until the sauce is thickened. Add the thickening. Transfer to the clay-pot. Garnish with scallion and red chili pepper.

魚香茄子順風煲

Sichuan Braised Eggplant and **Pork Ear**

份量：**4-6位用** ■製作時間：**10.5小時**

■工具：**7吋瓦煲1個（預熱）**

■ Serves: **4-6**

■ Preparation and Cooking Time: **10.5 hours**

■ Utensil: **1 clay-pot, 7-inch in diameter (pre-heated)**

材料：
新鮮或冰鮮豬耳2隻
茄子6兩（225克）
鹹魚茸½茶匙
葱度1條
紅椒絲¼隻

味料A：
凍白滷水1斤（600克）（參閱第123頁）

味料B：
豆瓣醬¼茶匙
柱侯醬⅙茶匙
砂糖½湯匙
雞粉¼湯匙
紹酒¼茶匙
清湯4兩（150克）
麻油⅙茶匙
胡椒粉⅒茶匙

芡汁：
生粉½茶匙＋清水2湯匙（調勻）

Ingredients:
2 pig ears
225g eggplant
½ tbsp salted fish, minced
1 stk scallion
¼ red chili pepper

Seasonings A:
600g white marinade (refer to p.123)

Seasonings B:
¼ tsp soy bean paste
⅙ tsp Chu Hau sauce
½ tbsp granulated sugar
¼ tbsp chicken powder
¼ tsp Shaoxing wine
150g broth
⅙ tsp sesame oil
⅒ tsp pepper

Thickening:
½ tsp cornstarch + 2 tbsp water

TIPS 偉師傅的專業指導

焯豬耳時，切勿蓋上鍋蓋，否則豬耳會不斷滲出油脂，失去爽脆口感。

Do not cover the saucepan while blanching the pig ears. Otherwise, the pig ears will sweat as oil is coming out. The meat will lose its crunchiness.

做法：

1. 把豬耳解凍，燒毛，洗乾淨後放入滾水裏焯30分鐘，取出泡冷水，放入味料A裏泡10小時，再切成1吋×½吋的長條，備用。
2. 茄子像梅花間竹般削外皮，切成3吋的小段，每段再切成6條長條形，洗好，用滾油炸，泡滾水，瀝乾。
3. 在鍋裏下油，炒香鹹魚茸，加入豆瓣醬和柱侯醬，炒香，放入茄子條和豬耳炒勻，灒紹酒，加入清湯和其他味料，煮至約剩下¼份清湯，倒下芡汁，轉放入瓦煲裏，加入葱度和紅椒絲便成。

Method:

1. Defrost the pig ears. Burn the hair. Wash the meat thoroughly and blanch for 30 minutes. Marinate with seasonings A for 10 hours. Cut them into julienne about 1 inch long x ½ inch thick. Set aside.
2. Zebra peel the eggplants. Cut into 3 inches thick. Cut each small pieces into 6 long pieces. Rinse them. Deep-fry. Boil the eggplants in hot water. Drain and dry.
3. Heat the wok. Sauté the salted fish. Add the soy bean paste, Chu Hau sauce, eggplants and pig ears. Sauté the mixture till fragrant. Add the wine, broth and other seasonings. When the broth is reduced to ¼ of its volume, add the thickening. Transfer to the clay-pot. Garnish with scallion and red chili pepper.

蝦醬豆腐豬耳煲

Braised Tofu and Pig Ear in Shrimp Paste Sauce

份量：**4-6位用** ■製作時間：**10.5小時**

■工具：**7吋瓦煲1個（預熱）**

■ Serves: **4-6**

■ Preparation and Cooking Time: **10.5 hours**

■ Utensil: **1 clay-pot, 7-inch in diameter (pre-heated)**

材料：
豬耳2隻
布包豆腐2件
蔥度1條
紅椒絲¼隻

味料A：
凍白滷水1斤（600克）（參閱第123頁）

味料B：
蒜茸2粒
蝦醬½茶匙
蠔油⅙茶匙
生抽⅛茶匙
砂糖½茶匙
雞粉½茶匙
麻油⅛茶匙
胡椒粉⅒茶匙
紹酒¼茶匙
清湯4兩（150克）

芡汁：
生粉½茶匙＋清水2湯匙（調勻）

Ingredients:
2 pig ears
2 pcs tofu
1 stk scallion, julienne
¼ red chili pepper, julienne

Seasonings A:
600g white marinade (refer to p.123)

Seasonings B:
2 cloves garlic, minced
½ tsp shrimp paste
⅙ tsp oyster sauce
⅛ tsp light soy sauce
½ tsp granulated sugar
½ tsp chicken powder
⅛ tsp sesame oil
⅒ tsp pepper
¼ tsp Shaoxing wine
150g broth

Thickening:
½ tsp cornstarch + 2 tbsp water

TIPS 偉師傅的專業指導

每款蝦醬，濃度不同，如覺得鹹味重，可加點砂糖。相反地，味道太淡可按喜好加蠔油或生抽。而喜歡吃辣者，則可加點豆瓣醬。

If you find the shrimp paste too salty, you can add some granulated sugar. Or you can add some oyster sauce or light soy sauce to enhance the flavor. If you prefer spicy taste, you can add some chili bean sauce.

做法：
1. 冰鮮豬耳解凍，燒去毛，清洗乾淨，焯30分鐘，取出泡冷水，加到味料A裏浸10小時，再取出切成1吋 × ½吋的長條形。
2. 把布包豆腐瀝乾，切成9塊，放溫油裏炸至金黃色。
3. 用油炒爆香蒜茸和蝦醬，加入豬耳和豆腐，快炒，潛紹酒，加入味料B炒勻，待汁液煮稠後，倒下芡汁放入煲裏，加入蔥度和紅椒絲便成。

Method:
1. Defrost the pig ears. Burn the hair. Wash thoroughly and blanch for 30 minutes. Marinate with seasonings A for 10 hours. Cut them into julienne of 1 inch long x ½ inch thick. Set aside.
2. Drain the tofu. Cut into 9 cubes. Fry until the tofu golden on both sides.
3. Heat the wok with oil. Sauté the garlic and shrimp paste. Add pig ears and tofu. Add wine and seasonings B. When the sauce is reduced, transfer the mixture to the clay-pot. Add the thickening. Garnish with scallion and red pepper.

口水雙脹

Braised Pork Knuckle and
Beef Shank with
Hot and Spicy Sauce

份量：**8-10位用** ■製作時間：**20小時**
■ Serves: **8-10**
■ Preparation and Cooking Time: **20 hours**

材料：

冰鮮加拿大豬肘2隻
冰鮮金錢牛脹2條
葱絲2條
紅椒絲1隻
炸花生碎2兩（75克）
熟芝麻1湯匙

味料：

凍白滷水3斤（1.8千克）（參閱第
123頁）

口水辣汁料：

砂糖8湯匙，精鹽2湯匙，雞粉
2湯匙，陳醋4湯匙，老抽1湯
匙，生抽1湯匙，麻油4湯匙，
麻醬2湯匙，魚露2湯匙，花生
醬2湯匙，辣椒油4湯匙，花椒
油1湯匙，川椒粉⅕湯匙，花椒
粉⅕湯匙，孜然粉⅕湯匙，清水
6湯匙

Ingredients:

2 pork shanks
2 beef shanks
2 stks scallion, julienne
1 red chili pepper, julienne
75g fried-peanut, chopped
1 tbsp sesame, sautéed

Seasonings:

1.8kg white marinade (refer to p.123)

The Hot and Spicy Sauce:

8 tbsp granulated sugar, 2 tbsp salt, 2 tbsp chicken powder, 4 tbsp mature vinegar, 1 tbsp dark soy sauce, 1 tbsp light soy sauce, 4 tbsp sesame oil, 2 tbsp sesame paste, 2 tbsp fish sauce, 2 tbsp peanut butter, 4 tbsp chili oil, 1 tbsp Chinese prickly ash oil, ⅕ tbsp Szechwan chili powder, ⅕ tbsp paprika, ⅕ tbsp cumin powder, 6 tbsp water

TIPS 偉師傅的專業指導

煲豬肘和牛脹時，勿蓋上煲蓋，否則肉會變軟了，不夠爽脆。

Leave the saucepan open while cooking the pork and beef shanks. Do not cover with a lid as the meat will lose its crunchiness.

做法

1. 把白滷水材料置煲中煮滾，轉慢火熬煮20分鐘，放涼後加入玫瑰露酒，備用。

2. 把口水辣汁混合，煮滾，放涼備用。

3. 將豬肘和牛脹解凍，放入滾水裏，用慢火煲50分鐘，熄火，原鍋浸50分鐘，取出，放在水喉下用慢流水沖洗10分鐘，用刀起出豬肘大骨，再用慢流水沖洗5分鐘，放涼凍。

4. 把豬肘和牛脹一同瀝乾，放入凍白滷水裏浸滷18小時。

5. 取出切薄片，淋上口水辣汁，撒上葱絲、紅椒絲、花生碎和熟芝麻便成。

Method:

1. Bring the white marinade to boil. Simmer for 20 minutes. Leave to cool. Add the wine. Set aside.

2. Add the hot and spicy sauce and bring to boil, set aside and leave to cool.

3. Defrost the pork and beef shanks. Simmer in hot water for 50 minutes. Turn off the fire and leave for 50 minutes. Wash under slow running water for 10 minutes. Bone the knuckle. Wash under slow running water for 5 minutes. Leave to cool.

4. Drain the knuckle and the beef shank. Marinate with cold white marinade for 18 hours.

5. Thinly slice the meat. Pour the hot and spicy sauce over the meat. Garnish with scallion, red chili pepper and sesame.

麻辣金肚絲

Shredded Ox Tripe in
Hot and Spicy Sauce

份量：**15位用** ■製作時間：**14小時**
■ Serves: **15**
■ Preparation and Cooking Time: **14 hours**

材料：
冰鮮金錢牛肚3個重約2斤（約1.2千克）

味料：
凍白滷水2斤（1.2千克）（參閱第123頁）

川辣醬汁：
雞粉1湯匙
辣油2湯匙
白醋2湯匙
老抽1湯匙
花椒油½湯匙
砂糖4湯匙
精鹽1湯匙
清水2湯匙
川椒粉¹⁄₁₀湯匙
孜然粉¹⁄₁₀湯匙
花椒粉¹⁄₁₀湯匙
麻油2湯匙

Ingredients:
3 frozen honeycomb ox tripe, 1.2kg

Seasonings:
1.2kg white marinade (refer to p.123)

Sichuan Chili Sauce:
1 tbsp chicken powder
2 tbsp chili oil
2 tbsp white vinegar
1 tbsp dark soy sauce
½ tbsp Chinese prickly ash oil
4 tbsp granulated sugar
1 tbsp salt
2 tbsp water
¹⁄₁₀ tbsp Sichuan chili powder
¹⁄₁₀ tbsp cumin powder
¹⁄₁₀ tbsp paprika
2 tbsp sesame oil

⊠ TIPS 偉師傅的專業指導

牛肚泡完冷水後，將牛肚中間剖開，取走裏面的白色肥油，否則切絲後就不美觀。

To have a good presentation, cut the tripe in the middle and trim the fat after cold water bath.

做法：
1. 把白滷水材料置煲中煮滾，轉慢火熬煮20分鐘，放涼後加入玫瑰露酒，備用。
2. 把川辣醬汁材料混和，煮滾，放涼，備用。
3. 牛肚解凍，放入滾水裏煮70-80分鐘，取出泡冷水後，放進凍白滷水裏浸滷12小時。
4. 將牛肚取出，切絲，與川辣醬汁拌勻便成。

Method:
1. Bring the white marinade to boil. Simmer for 20 minutes. Leave to cool. Add the wine. Set aside.
2. Add the Sichuan chili sauce and bring to boil. Set aside. Leave to cool.
3. Defrost the ox tripe. Cook in boiling water for 70-80 minutes. Drain. Cold water bath. Drain again. Marinate with the white marinade for 12 hours.
4. Cut the ox tripe into julienne. Toss with the Sichuan chili sauce. Serve.

泰汁金肚絲

Shredded Ox Tripe in Thai Sauce

份量：**15位用** ■製作時間：**14小時**
■ Serves: **15**
■ Preparation and Cooking Time: **14 hours**

材料：

冰鮮金錢牛肚3個重約2斤（1.2千克）

味料：

凍白滷水2斤（1.2千克）（參閱第123頁）

泰式醬汁：

泰式雞醬6湯匙
魚露2湯匙
太白司高辣椒汁¼湯匙
麻油2湯匙
甘筍¼條，切絲
洋葱½個，切絲
指天椒5隻，切絲
乾葱肉5粒，切絲
香芹2兩（75克），切絲
香茅2條，切絲
鮮薄荷葉5片，切絲
泰國青檸汁2個

Ingredients:
3 honeycomb ox tripe, 1.2kg

Seasonings:
1.2kg white marinade (refer to p.123)

Thai Sauce:
6 tbsp Thai sweet chili sauce
2 tbsp fish sauce
¼ tbsp Tabasco sauce
2 tbsp sesame oil
¼ carrot, julienne
½ onion julienne
5 red cluster peppers, shredded
5 cloves shallot, shredded
75g parsley, shredded
2 stks lemon grass, shredded
5 fresh mint leaves, shredded
2 Kaffir lime s from Thai

↘ TIPS 偉師傅的專業指導

煮牛肚時要注意牛肚的品質。小牛肚只需煮50分鐘便可；老牛肚就差不多要煮90分鐘了。

We need to pay attention to the quality of the tripe. It needs 50 minutes to cook calf's tripe and 90 minutes for ox tripe.

做法：

1. 把白滷小材料直煲中煮滾，用慢火熬煮20分鐘，放涼後加入玫瑰露酒，備用。

2. 把泰式醬汁調和，備用。

3. 牛肚解凍，放入滾水裏煮70-80分鐘，取出，泡冷水，放進凍白滷水裏浸滷12小時。

4. 取出牛肚，切絲，拌入泰式醬汁便成。

Method:

1. Bring the white marinade to boil. Simmer for 20 minutes. Leave to cool. Add the wine. Set aside.

2. Mix the Thai sauce. Set aside.

3. Defrost the ox tripe. Cook in boiling water and for 70-80 minutes. Take out and immerse in the cold water. Drain. Marinate with the white marinade for 12 hours.

4. Cut the ox tripe into julienne. Toss with the Thai sauce.

五彩金肚絲

Shredded Ox Tripe with
Vegetables

份量：**15位用** ■製作時間：**14小時**
■Serves: **15**
■Preparation and Cooking Time: **14 hours**

材料：
冰鮮金錢牛肚3個重約2斤（1.2千克）
西芹半棵，切絲
甘筍1條，切絲，汆水，泡冷水
木耳2兩（75克），切絲，汆水，泡冷水

味料：
凍白滷水2斤（1.2千克）（參閱第123頁）

拌味料：
砂糖1湯匙
精鹽½湯匙
雞粉½湯匙
麻油1湯匙
豆瓣醬½湯匙
熟芝麻½湯匙
蕎頭10粒，切絲
酸薑10片，切絲
京葱1條，切絲

Ingredients:
3 honeycomb ox tripe, 1.2kg
½ stk celery, julienned
1 carrot, blanched and julienned
75g wooden fungus, blanched and julienned

Seasonings:
1.2kg white marinade (refer to p.123)

Condiment:
1 tbsp granulated sugar
½ tbsp salt
½ tbsp chicken powder
1 tbsp sesame oil
½ tbsp broad bean paste
½ tbsp sesame, sautéed
10 cloves pickled shallot, shredded
10 slices pickled ginger, shredded
1 stk Peking scallion, shredded

⬊ TIPS 偉師傅的專業指導

煲牛肚時，可加入1湯匙豆瓣醬，能辟除冰鮮味和腥味，還會令色澤更美觀。

To remove the frozen food and fishy smell, add one tbsp of chili bean sauce into the water while cooking the ox tripe. This can also add the color.

做法：

1. 將牛肚解凍，放入滾水裏煮70-80分鐘，取出，泡冷水。

2. 把牛肚放進凍白滷水裏浸滷12小時。

3. 將牛肚取出，切絲，加入全部材料和拌味料拌勻，便成。

Method:

1. Defrost the ox tripe. Cook in boiling water for 70-80 minutes. Take out and immerse into cold water.

2. Marinate the ox tripe with white marinade for 12 hours.

3. Cut the ox tripe into julienne. Toss with condiment and other ingredients.

燒烤醬辣汁羊腩煲

Braised Lamb Brisket in Chilli BBQ Sauce

份量：**10位用** ■製作時間：**1小時40分鐘**

■ Serves: **10**

■ Preparation and Cooking Time: **1 hour 40 minutes**

材料：
新鮮黑草羊腩2斤（1.2千克）
去皮馬蹄8兩（300克）
甘筍1棵
葱2條，切段
紅椒1隻，切角

味料A：
糯米酒2湯匙
薑粒3兩（113克）
乾葱頭粒5粒
蒜肉6粒
檸檬葉2片
果皮½個
草果2粒
香葉6片
燒烤醬2湯匙（參閱第121頁）
豆瓣醬1湯匙
紅椒粉¼湯匙
太白司高辣椒汁1湯匙

味料B：
胡椒粉½茶匙
清湯4斤（2.4千克）
精鹽1茶匙
冰糖1兩（38克）
雞粉1茶匙

Ingredients:
1.2kg mutton
300g water-chestnut, peeled
1 carrot
2 stks scallions, sectioned
1 red chili pepper, cut diagonally

Seasonings A:
2 tbsp glutinous wine
113g ginger, diced
5 cloves shallot
6 cloves garlic
2 Kaffir lime leaves
½ dried tangerine peel
2 brown cardamom pods
6 bay leaves
2 tbsp BBQ sauce (refer to p.121)
1 tbsp broad bean paste
¼ tbsp paprika
1 tbsp Tabasco sauce

Seasonings B:
½ tsp pepper
2.4kg broth
1 tsp salt
38g rock sugar
1 tsp chicken powder

◣ TIPS 偉師傅的專業指導

炆羊腩要看選料，如老羊的話，會很難炆軟，口感粗糙，可多加清水和冰糖以慢火再煮60分鐘。

It takes longer time to braise the meat of old sheep to tender. We need to add more water and rock sugar and 1 more hour to braise the meat.

做法：

1. 把羊腩斬成約2吋 × 2吋的方塊，然後放入白鑊裏炒乾，灒入糯米酒炒片刻，加入滾水煮沸，取出瀝乾，再次燒熱鍋，把羊腩件炒乾。

2. 甘筍去皮，切斜角。馬蹄放入滾水裏汆水，泡冷水。

3. 燒熱油鍋，爆香味料A，加入羊腩炒勻，再加入味料B煮滾，放在已墊竹笪的瓦煲裏，防止物料黏底，並以慢火煮1小時，放入甘筍和馬蹄繼續煮30分鐘，取走檸檬葉、果皮、草果和香葉，撒上葱度和紅椒絲便成。

Method:

1. Cut the mutton into 2 inches x 2 inches. Sauté the meat. Add glutinous wine and sauté a while. Add water and bring to boil. Drain and dry. Heat the wok and sauté the meat again until all the cooking liquid is reduced.

2. Peel the carrot and cut diagonally. Blanch the water chestnut. Immerse it into cold water.

3. Heat the wok with oil and then sauté seasonings A till fragrant. Add the meat. Add seasonings B and bring to boil. Transfer to a clay-pot with a pieces of bamboo mat in it. The mat is used to prevent food from sticking to the pot. Braise the meat for an hour. Add carrots and water chestnuts. Braise for another 30 minutes. Take away the lemon grass, dried tangerine peel, brown cardamom pods and bay leaves. Garnish with scallion and red chili pepper.

香芒鵝片
Sliced Goose with **Mango**

份量：**4-6位用** ■製作時間：**5分鐘**

■ Serves: 4-6

■ Preparation and Cooking Time: **5 minutes**

材料：
熟燒鵝胸1個
呂宋香芒2個
竹葉2片

Ingredients:
1 pcs goose breast, roasted
2 mangoes
2 bamboo leaves

◥ TIPS 偉師傅的專業指導

1. 處埋香芒鵝片時，一定
   要先切好芒果，再切其
   他，切勿打亂次序。否
   則鵝胸變黃；芒果變
   黑，賣相就很不美觀。

2. 竹葉在山貨舖有售。

1. The procedure is very
   important. Make sure
   to slice the mango and
   prepare the juice first.
   Otherwise, the color of
   the goose breast will
   turn yellow and the
   mango will turn black.
   The food presentation
   will not look good.

2. We can buy bamboo
   leaves in local Chinese
   grocery stores.

做法：

1. 將熟燒鵝胸去骨，將鵝肉連皮切成12件。

2. 把一個芒果削皮去核，攪拌成芒果汁；另一個芒果橫切成3份，去皮，斜斜切成七小塊。

3. 用清水把竹葉洗淨，抹乾，分放在兩碟上，然後按序排放1塊芒果片和鵝片，如是者共排放7件芒果片和6件鵝片，並把芒果汁慢慢注入兩邊便成。

Method:

1. Debone the roast goose breast. Cut the meat with the skin on into 12 slices.

2. Peel a mango, remove the shell and blended into juice. Cut the remains into 3 small pieces. Peel the skin. Further cut diagonally into seven small pieces.

3. Wash and dry the bamboo leaves. Put one each on a plate. Arrange the mango slices and goose breast slices on the plate. Each plate will have 7 mango slices and 6 goose breast slices. Put the mango juice aside.

鮑汁鵝掌翼煲

Braised Goose Web and Wing in **Abalone Sauce**

份量：**4-6位用** ▓製作時間：**11小時**

▓工具：**7吋瓦煲1個（預熱）**

▓ Serves: **4-6**

▓ Preparation and Cooking Time: **11 hours**

▓ Utensil: **1 clay-pot, 7-inch in diameter (pre-heated)**

材料：
冰鮮鵝掌翼各3隻
葱度1條
紅椒絲¼隻

味料A：
凍白滷水2斤（1.2千克）（參閱第123頁）

味料B：
清湯10兩（375克）
蠔油1湯匙
雞粉½湯匙
砂糖½湯匙
老抽¼茶匙
麻油¼茶匙
生薑2片
蒜肉1粒
洋葱¼個
八角1粒

芡汁：
生粉½茶匙＋清水2湯匙（調勻）

Ingredients:
3 goose webs and 3 goose wings
1 stk scallion
¼ red chili pepper, shredded

Seasonings A:
1.2kg white marinade (refer to p.123)

Seasonings B:
375g broth
1 tbsp oyster sauce
½ tbsp chicken powder
½ tbsp granulated sugar
¼ tsp dark soy sauce
¼ tsp sesame oil
2 slices ginger
1 clove garlic
¼ onion
1 star anise

Thickening:
½ tsp cornstarch + 2 tbsp water

◥ TIPS 偉師傅的專業指導

清洗鵝掌時，要將掌上的黑色物質用刷子刷走，鵝翼也要徹底去毛，否則就不但不衛生，也極不美觀。

For the sake of good health and good presentation, make sure to wash and scrub the goose webs thoroughly to make them look good. It is also important to remove all the hairs on the wings too.

做法：
1. 把冰鮮鵝掌、鵝翼解凍，清洗乾淨後放入滾水裏用慢火焯煮10分鐘，熄火浸30分鐘，取出泡冷水。
2. 鵝掌翼放涼後，加入味料A裏浸10小時，取出，把鵝掌切開3份，翼切開4份。
3. 燒熱鍋，下少量油，爆香生薑、蒜肉、洋葱和八角後，加入味料B，改用慢火煮5分鐘，取出材料，倒下芡汁，將掌翼加入煮1分鐘，轉放已預熱的瓦煲裏再煮1分鐘。
4. 加入葱度和紅椒絲後蓋上蓋，便成。

Method:
1. Defrost the goose webs and wings. Wash. Simmer in hot water for 10 minutes. Turn off the fire and soak them in hot water for 30 minutes. Take out and immerse it into cold water bath.
2. Marinate the webs and wings in seasonings A for 10 hours. Chop each web into 3 small pieces and each wing into 4 small pieces.
3. Heat the wok with oil. Sauté the ginger, garlic, onion and star anise until fragrant. Add seasonings B and simmer for 5 minutes. Drain. Thicken the sauce. Add the webs and wings. Cook for 1 minute. Transfer the mixture to the preheated clay-pot and cook for 1 minute again.
4. Garnish with scallion and red chili pepper. Cover with a lid. Serve.

柱侯鵝掌翼煲

Goose Web and Wing in Chu Hau Sauce

份量：**4-6位用** ■製作時間：**11小時**

■工具：**7吋瓦煲1個（預熱）**

■ Serves: **4-6**

■ Preparation and Cooking Time: **11 hours**

■ Utensil: **1 clay-pot, 7-inch in diameter (pre-heated)**

材料：

冰鮮鵝掌翼各3隻
蔥度1條
紅椒絲¼隻

味料A：

凍白滷水2斤（1.2千克）（參閱第123頁）

味料B：

清湯10兩（375克）
雞粉½湯匙
砂糖½湯匙
麻油¼茶匙
老抽¼茶匙
蠔油¼茶匙
柱侯醬1湯匙
生薑2片
八角1粒
蒜肉1粒
洋蔥¼個

芡汁：

生粉½茶匙＋清水2湯匙（調勻）

Ingredients:

3 goose webs and 3 goose wings
1 stk scallion
¼ red chili pepper

Seasonings A:

1.2kg white marinade (refer to p.123)

Seasonings B:

375g broth
½ tbsp chicken powder
½ tbsp granulated sugar
¼ tsp sesame oil
¼ tsp dark soy sauce
¼ tsp oyster sauce
1 tbsp Chu Hau paste
2 slices ginger
1 star anise
1 clove garlic
¼ onion

Thickening:

½ tsp cornstarch + 4 tbsp water

⊠ TIPS 偉師傅的專業指導

煮鵝掌、翼時，味料B的份量必須要剛好能蓋過材料。

Make sure seasonings B has enough liquid to cover the goose webs and wings during the simmering process.

特式燒滷

凉菜小食

常用香料和醬料

做法：

1. 把冰鮮鵝掌、翼解凍後，清洗乾淨，放入滾水裏用慢火焯煮10分鐘，熄火，浸30分鐘，取出，泡冷水。

2. 鵝掌、翼放涼後，加入味A料裏浸10小時，取出，用刀把鵝掌切開3份，鵝翼切開4份。

3. 在鍋裏放油，爆香生薑、蒜肉、洋蔥、八角和柱侯醬，加入味B料，轉慢火煮5分鐘，取走材料，倒下芡汁，另加入鵝掌、翼，用慢火煮1分鐘，熄火，轉放瓦煲裏再煮1分鐘。

4. 加入蔥度和紅椒絲，蓋上鍋蓋，便成。

Method:

1. Defrost the goose webs and wings. Wash. Simmer in hot water for 10 minutes. Turn off the fire and immerse in the hot water for 30 minutes. Immerse it into cold water bath.

2. Marinate the cold webs and wings in seasonings A for 10 hours. Chop each web into 3 small pieces and each wing into 4 small pieces.

3. Heat the wok with oil. Sauté the ginger, garlic, onion, star anise and Chu Hau sauce until fragrant. Add seasonings B and simmer for 5 minutes. Drain. Thicken the sauce. Add the webs and wings. Cook for 1 minute. Turn off the fire. Transfer the mixture to the pre-heated clay-pot and cook for 1 minute again.

4. Garnish with scallion and red chili pepper. Cover with a lid. Serve.

南京鹽水鴨

Nanjing Simmered Duck in
Salted Broth

份量：**10-12位用** ▊製作時間：**3小時**
▊ Serves: **10-12**
▊ Preparation and Cooking Time: **3 hours**

材料：
冰鮮瘦米鴨1隻重約3½-4斤
（2.1-2.4千克）
葱絲1棵
紅椒絲½隻

味料A：
薑10片
京葱2條，切段
蒜肉6粒
紅椒1隻

味料B：
清水10斤（6千克）
甘草3片
草果3粒
果皮¼個
花椒10粒
香葉10片
丁香2粒
白胡椒½湯匙
八角2粒

味料C：
精鹽4兩（150克）
雞粉3湯匙
魚露3湯匙
玫瑰露酒3湯匙

Ingredients:
1 duck (about 2.1 - 2.4kg)
1 stk scallion, shredded
½ red chili pepper, shredded

Seasonings A:
10 slices ginger
2 stk Peking scallion, sectioned
6 cloves garlic
1 red chili pepper

Seasonings B:
6 kg water
3 liquorices
3 brown cardamom pods
¼ dried tangerine peel
10 Sichuen pepper
10 bay leaves
2 cloves
½ tbsp white pepper
2 star anise

Seasonings C:
150g salt
3 tbsp chicken powder
3 tbsp fish sauce
3 tbsp rose wine

↘ TIPS 偉師傅的專業指導

1. 要想知道鴨是否已熟，可用竹籤或鋼針在鴨胸最厚的部份刺入，然後拔出，如果流出的肉汁是紅色則表示鴨肉未熟；如汁液清澈代表鴨肉已熟。

2. 不想用鴨油，用生油也可。

3. 鴨的烹調時間可按鴨的老嫩程度做調節。

1. To test if the duck is cooked or not, prick a needle into the thickest part of the duck breast. If the juice comes out is red in color, it means the duck is still undercooked. If the juice is crystal clear, it means the duck is ready to serve.

2. Peanut oil can substitute duck fat.

3. Cooking time depends on the duck.

做法：
1. 挖出米鴨的鴨油、肺和喉管，洗淨，瀝乾，備用。
2. 燒熱鍋，下鴨油爆香味料A，加入味料B裏滾煮，改用慢火熬煮30分鐘，再加入味料C，煮至精鹽完全溶解。
3. 在味料裏，加墊一竹笪，放入米鴨，煮沸後轉用慢火煮40-50分鐘。
4. 取出，稍微放涼後切塊，鴨面放葱絲和紅椒絲便成。

Method:
1. Trim all duck fat. Remove the lung and the throat. Wash. Drain. Set aside.
2. Heat the wok with duck's fat. Sauté seasonings A till fragrant. Add seasonings B and bring to boil. Simmer for 30 minutes. Add seasoning C, stir until all the salt is dissolved.
3. Add a bamboo mat in the liquid. Put in the duck. When the mixture brings to boil, reduce the heat. Simmer for 40-50 minutes.
4. Take out the duck and leave to cool. Chop the duck. Garnish with scallion and red chili pepper.

堂剪妙齡鴨

Roasted **Duckling**

份量：**4-6位用** ■製作時間：**3小時**
■ Serves: **4-6**
■ Preparation and Cooking Time: **3 hours**

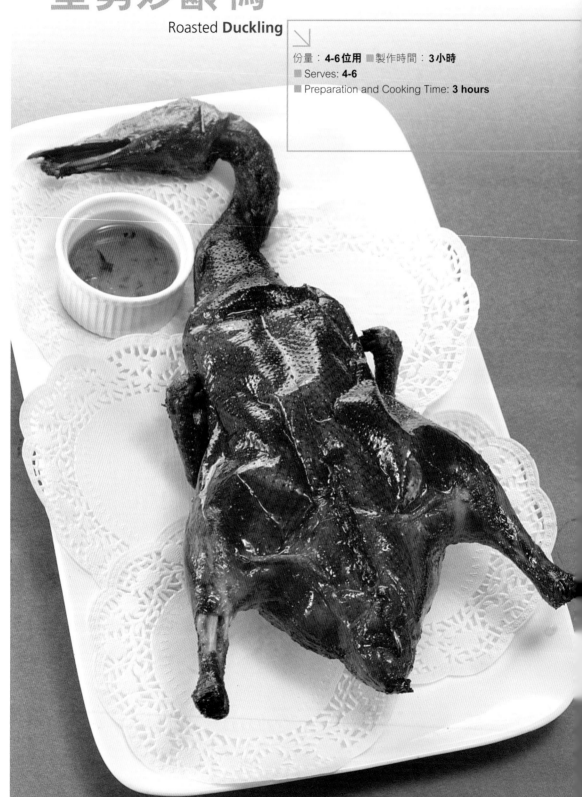

材料：
冰鮮小米鴨1隻重約2¼-2½斤（1.35-1.5千克）
生油2碗（400克）

味料：
精鹽1茶匙
雞粉½茶匙
砂糖1茶匙
五香粉1/10茶匙
燒烤醬2茶匙（請參閱第121頁）
玫瑰露酒1茶匙
炸蒜茸1茶匙

上皮：
燒雞水適量（參閱第123頁）

蘸汁：
冰花梅醬3湯匙

Ingredients:
1 duck (1.35-1.5kg)
400g peanut oil

Seasonings:
1 tsp salt
½ tsp chicken powder
1 tsp granulated sugar
1/10 tsp five spices powder
2 tsp BBQ sauce (refer to p.121)
1 tsp rose wine
1 tsp garlic, deep fried

Brush-On Glaze:
Brine for Roasting Chicken, appropriate (refer to p.123)

Dipping:
3 tbsp plum sauce

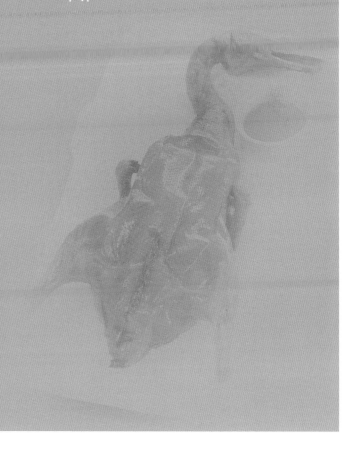

堂剪妙齡鴨
Roasted Duckling

做法：

1. 將冰鮮米鴨解凍，挖去油脂、肺和喉管，洗淨，瀝乾，把味料均勻地抹在鴨胸腔裏，再用鋼針縫口。

2. 把鴨身洗淨，並從鴨頸位置吹入空氣，使鴨胸膨脹。汆水，取出，在鴨皮抹上上皮料，掛起風乾2小時。

3. 焗爐預熱，把米鴨放在烘烤盆裏的鐵架上，蓋上錫紙，放入焗爐以140℃火烘烤30-35分鐘，取出待涼。再拉油，瀝乾，上碟。

4. 上桌時按下面的方法剪開，蘸冰花梅醬伴食。

Method:

1. Defrost the duck. Trim all duck fat. Remove the lung and the throat. Wash. Drain. Set aside. Season the duck breast and the cavity. Sew the duck with a cooking needle.

2. Wash the duck. Inject air into the cavity through the neck like a balloon. Blanch the duck. Blast with bush-on glaze. Hang the duck and air-dry for 2 hours.

3. Preheat the oven. Put the duck inside the oven. Cover it with aluminum foil. Roast at 140℃ for 30 -35 minutes. Take it out and leave to cool. Poach the duck with oil till cooked. Drain and serve.

4. Cut the duck with scissors as steps shown below. Serve with plum sauce.

◥ TIPS 偉師傅的專業指導

1. 食用時，拔出鋼針，鴨尾向下，把鴨汁小心地注入碟裏（圖1）。

2. 鴨胸向上，用骨剪沿鴨尾向上剪開鴨胸，最後到頸部（圖2）。

3. 翻過鴨身，成琵琶形，再將鴨身剪十字形，除頭頸外分成4份，再將每份開6件（圖3-5）。鴨皮切勿黏上汁液，否則會不酥脆。

1. Remove the needle before serving. Pour the gravy carefully into a plate (fig 1).

2. With the duck breast facing up, cut along the rare end to the neck (fig 2).

3. Put the duck on the plate with its back down, just like a mandolin. Except the neck of the duck, divide the duck from the middle into 4 equal parts. And each part further divided into 6 small pieces (fig 3-5). In order to keep the skin crispy, avoid the gravy in touch with the skin.

剪鴨方法 Cutting the Duck

1	2
3	4
5	

1. 先取出鐵針,倒出汁液。
2. 用剪刀沿鴨胸剪成兩半。
3. 把鴨剪成四塊。
4. 按不同部位剪成小塊。

1. Remove the steel needle and throw out the grave from the duck.
2. Cut the duck into a half with a pair of scissor along it's breast.
3. Cut the duck into 4 pieces with a pair of scissor.
4. According to different parts of the duck, cut into small pieces.

海蜇火鴨絲

Shredded Roasted Duck
with **Jelly Fish**

份量：**6-8位用** ■製作時間：**5分鐘**

■ Serves: **6-8**

■ Preparation and Cooking Time: **5 minutes**

材料：

熟火鴨¼隻（要上肢部份，做法參
閱第50頁）

海蜇皮1斤（600克）

白醋2湯匙

熟芝麻1茶匙

紅椒絲½隻

現成五柳料4兩（150克）

味料：

砂糖1湯匙

精鹽½湯匙

雞粉½湯匙

麻油1湯匙

豆瓣醬½湯匙

Ingredients:

¼ pcs roasted duck (upper part, refer to p.50)

600g jelly fish

2 tbsp white vinegar

1 tbsp sesame, sautéed

½ red chili pepper, shredded

150g assorted pickles

Seasonings:

1 tbsp granulated sugar

½ tbsp salt

½ tbsp chicken powder

1 tbsp sesame oil

½ tbsp broad bean paste

TIPS 偉師傅的專業指導

五柳料在醬料舖有售。它
是用蕎頭（蕎頭）、蘇薑（紅
薑）、嫩子薑（白酸薑）、木
瓜絲（瓜英）和甘荀絲（錦
菜）等五種材料，加入白醋
和砂糖醃製而成。

We can buy assorted pickles from grocery store which sell sauces and condiments. Assorted pickles are a mix of pickled shallot, pickled ginger, white sour ginger, shredded melon and carrot. They are marinated with granulated sugar and white vinegar.

做法：

1. 把海蜇皮捲成筒形，切成約1枝普通筷子的厚度，用清水慢慢地沖洗30分鐘，取出瀝乾。

2. 在海蜇裏加入白醋，拌勻，注入2斤（1.2千克）滾水，快速拌勻，倒去熱水，泡冷水，再放在水喉下以慢速流水沖30分鐘，讓其脹發。

3. 瀝乾，加入味料拌勻，醃10分鐘，取出瀝乾，再加入五柳料和白芝麻拌勻，置放碟中。

4. 把火鴨去骨，鴨肉切絲放在海蜇上面，加點紅椒絲點綴便成。

Method:

1. Roll jelly fish into a cylinder. Cut into slice which each slice is as thick as a chopstick. Rinse the jelly fish slowly under running water for 30 minutes. Drain and dry. Set aside.

2. Toss the jelly fish with white vinegar. Add 1200g hot boiling water. Stir. Drain. Cold water bath. Rinse slowly under running water for 30 minutes. The jelly fish will swell.

3. Drain and dry. Marinate with seasonings for 10 minutes. Drain and dry again. Toss with assorted pickles and sesame. Put them in a plate.

4. Bone the duck. Shred it. Put the meat over the jelly fish. Garnish with red chili pepper. Serve.

沙爹燒鴨

Roasted Duck with
Satay Sauce

份量：**10-12位用** ■製作時間：**35-40分鐘**
■ Serves: **10-12**
■ Preparation and Cooking Time: **35-40 minutes**

材料：
冰鮮米鴨1隻重約4斤 - 4½斤
（2.4 - 2.7千克）

味料A：
蒜茸4粒
乾葱茸2粒
陳皮茸1棵
薑茸2片
沙爹醬4湯匙

味料B：
精鹽3湯匙
砂糖2湯匙
雞粉1湯匙
五香粉¼湯匙
玫瑰露酒1湯匙
芫荽1棵
八角3粒
葱度1棵

上皮：
白醋2兩（75克）
浙醋1湯匙
麥芽糖2兩（75克）
紹酒1湯匙
清水1兩（38克）

Ingredients:
1 duck (about 2.4-2.7kg)

Seasonings A:
4 cloves garlic, minced
2 cloves shallot, minced
1 dried tangerine peel, fine chopped
2 slices ginger, fine chopped
4 tbsp satay sauce

Seasonings B:
3 tbsp salt
2 tbsp granulated sugar
1 tbsp chicken powder
¼ tbsp five spices powder
1 tbsp rose wine
1 stk parsley
3 star anise
1 stk scallion

Brush-On Glaze:
75g white vinegar
1 tbsp red vinegar
75g maltose
1 tbsp Shaoxing wine
38g water

TIPS 偉師傅的專業指導

一般家用式焗爐比專業燒鴨爐的火力弱，熱力相對不夠均勻，所以要等燒鴨烘烤熟後才拉油，色澤會較平均，效果也最理想。

The industrial-use oven generally has a better cooking result than domestic oven due to the heating power. In order for the domestic oven to achieve a better result, poach the duck with oil after roasting in the oven. The color of the duck will be even and beautiful.

做法：

1. 燒熱鍋，爆香味料A，備用。將味料B和上皮料分別混合，備用。

2. 把米鴨挖去油脂、肺和喉管，洗淨瀝乾，均勻地抹上味料A和味料B，用鋼針縫口，洗淨。

3. 把米鴨放進滾水裏燙皮，取出，均勻地在鴨皮抹上上皮料，掛起，風乾4小時。

4. 焗爐預熱，轉中火約180℃，放入米鴨，蓋上錫紙，烤約35-40分鐘取出，食用時才拉油。

Method:

1. Heat the wok and sauté seasonings A to fragrant. Set aside. Mix seasonings B and the brush-on glaze individually. Set aside.

2. Trim all fat from the duck. Remove the lung and throat. Wash and drain. Evenly blast the duck with seasonings A and B. Sew the duck with a cooking needle. Wash the duck thoroughly.

3. Blanch the duck to soften the skin. Take it out. Blast the duck with brush-on glaze. Hang the duck and air dry for 4 hours.

4. Preheat the oven. Cover the duck with aluminum foil. Roast the duck at 180℃ for 35-40 minutes. Poach the duck with hot oil before serving.

圍村南乳鴨

Braised Duckling with Red Sauce

份量：**8-10位用** ▦製作時間：**50分鐘**
Serves: **8-10**
▪Preparation and Cooking Time: **50 minutes**

材料：
冰鮮瘦米鴨1隻重約3½-4斤
（2.1-2.4千克）

味料A：
陳皮2片，剁茸
蒜茸6粒
乾葱茸3粒
南乳1大件，壓爛成茸
蠔油1湯匙

味料B：
八角1粒
甘草3片
草果2粒
清水5斤（3千克）
精鹽3湯匙
冰糖4兩（150克）
雞粉2湯匙
紹酒1湯匙
玫瑰露酒1湯匙

Ingredients:
1 chilled lean duck (about 2.1-2.4kg)

Seasonings A:
2 dried tangerine peels, fine chopped
6 cloves garlic, minced
2 cloves shallot, minced
1 pcs fermented red beancurd, crushed
1 tbsp oyster sauce

Seasonings B:
1 star anise
3 pcs liquorices
2 brown cardamom pods
3 kg water
3 tbsp salt
150g rock sugar
2 tbsp chicken power
1 tbsp Shaoxing wine
1 tbsp rose wine

⟋ TIPS 偉師傅的專業指導

1. 煮鴨時，緊記先墊竹笪，可防鴨皮黏底，弄破鴨的表面。

2. 煮鴨的火候要控制恰當，才能讓鴨汁液收乾。

3. 確保米鴨一定要熟透，若水份不足，可在中途酌量添水，令鴨肉熟透、變軟。當汁液開始變稠，會黏貼在鴨肉上，熱力便很難滲入，鴨肉也較難熟。

1. To prevent the duck stick to the wok and spoiled the beauty, put a bamboo mat on the bottom of the wok first.

2. Alert of the fire in order to reduce the gravy of the duck.

3. If there is not enough water or the gravy is too thick, it is difficult to cook the duck to tender. Therefore, make sure to have enough water during the cooking process. Add water if necessary.

做法：
1. 把米鴨挖去油脂、肺和喉管，洗淨，瀝乾。
2. 燒熱油鍋，爆香味料A，放入米鴨，煎至兩面金黃，散發香味。
3. 加入味料B煮滾，轉慢火烹煮40分鐘。
4. 取出米鴨稍微放涼、切塊，原汁淋在鴨件上，趁熱享用。

Method:
1. Trim all fat from the duck. Remove the lung and throat. Wash and drain.
2. Heat oil in wok and sauté seasonings A to fragrant. Panfry the duck until golden brown and fragrant.
3. Add seasonings B and bring to a boil. Reduce the heat and simmer for 40 minutes.
4. Take out the duck. Chop later. Pour the gravy over the duck. Serve hot.

鮮果煙鴨胸

Smoked Duck Breast with **Guava**

份量：**4-6位用** ▍製作時間：**15-20分鐘**
▍Serves: **4-6**
▍Preparation and Cooking Time: **15-20 minutes**

材料：
冰鮮煙鴨胸1個
大石榴1個
竹葉2塊

石榴蘸汁料：
石榴 ½ 個

Ingredients:
1 pcs frozen smoked duck breast
1 pc guava
2 bamboo leaves

Pipping sauce:
Guava, juicy

◥ TIPS 偉師傅的專業指導

1. 鮮果可配富士蘋果、哈蜜瓜、啤梨、奇異果、新鮮菠蘿等。
2. 蘸食醬料可改用冰花梅醬或冰梅汁。

1. Any kinds of fruit we can use, such as Fuji apple, cantaloupe, pear, kiwi and pineapple, etc.
2. For dipping sauce, we can use plum paste or plum sauce.

做法：

1. 煙鴨胸沖洗，放入焗爐以約180℃中火烘烤15-20分鐘，取出，稍涼後切成12片薄片。
2. 石榴去核後½個搾汁備用，½個斜刀切成14件。
3. 用清水把竹葉洗淨，抹乾，分放在兩碟上。
4. 將7片石榴片、6片煙鴨胸片梅花間竹地排放在竹葉上。
5. 把石榴汁慢慢注在材料兩邊便成。也可用器皿盛載另上。

Method:

1. Wash the duck breast. Roast it in oven at 180℃ for 15-20 minutes. Remove. Cool for a while. Cut into 12 slices.
2. Core the guava and cut into a half. One part is for blending and the other part is for slicing. Leave for use.
3. Wash and dry the bamboo leaves. Put one each on a plate.
4. Evenly zebra place the guava and duck meat on the plates.
5. Serve with plum sauce.

湛江鹹香雞

Zhanjiang
Savory Chicken

份量：**6-8位用** ■製作時間：**1小時40分鐘**
■ Serves: **6-8**
■ Preparation and Cooking Time: **1 hour 40 minutes**

材料：
光雞1隻重約2½斤（1.5千克）
蔥絲1棵
紅椒絲½隻

輔料：
火腿1兩（38克），剁茸
蝦米1兩（38克），剁茸
蒜6粒，去皮剁茸
乾蔥3粒，剁茸
薑3片，切粒

味料A：
清水5斤（3千克）
甘草3片
草果3粒
沙薑6粒
香葉10片
丁香1粒
黃枝子2粒

味料B：
精鹽10兩（375克）
雞粉2湯匙
麥芽糖1湯匙
玫瑰露酒1湯匙
魚露1湯匙

Ingredients:
1 chicken (about 1.5kg)
1 stk scallion
½ red chili pepper, shredded

Minor Ingredients:
38g ham, fine chopped
38g dried shrimp, fine chopped
6 cloves garlic, peeled and minced
3 cloves shallot, minced
3 slices ginger, diced

Seasonings A:
3 kg water
3 licorices
3 black cardamom pods
6 zedoary
10 bay leaves
1 clove
2 cape jasmine fruit

Seasonings B:
375g salt
2 tbsp chicken powder
1 tbsp maltose
1 tbsp rose wine
1 tbsp fish sauce

◤ TIPS 偉師傅的專業指導

如果浸味汁份量不夠，而不能完全蓋過雞，可把雞胸切開兩半，做成琵琶形才泡汁，效果相若，但浸滷時間應縮減10分鐘。

If the seasonings is not enough to cover the whole chicken, cut the chicken into two pieces like a mandolin shape. It tastes the same. However, you need to reduce the cooking time by 10 minutes.

做法：
1. 把光雞挖去油脂、肺、喉管，洗淨。
2. 燒熱鍋，用少量油爆香輔料，加入味料A煮滾，轉慢火熬煮30分鐘，放入味料B，待鹽溶解。
3. 把雞放入味料裏煮滾，等雞裏充滿水份後撈起瀝乾，如是者做5次，熄火，加蓋，把雞留在汁裏浸30分鐘。
4. 取出雞，稍涼後斬件，加入蔥絲和紅椒絲點綴便成。

Method:
1. Trim all fat from the chicken. Remove the lung and throat. Wash and drain.
2. Heat the wok with oil. Sauté seasonings A to fragrant. Reduce the heat and simmer for 30 minutes. Add seasonings B until the salt is dissolved.
3. Cook in the seasonings and bring to boil. Drain and dry when the chicken is soaked with liquid. Repeat 5 times. Turn off the fire and leave the chicken in the liquid for 30 minutes.
4. Take out the chicken. Leave for a while. Chop the chicken. Garnish with scallion and red chili pepper.

葱花麻油雞

Steamed Chicken in
Sesame Oil

份量：**6-8位用** ■製作時間：**3小時**

■ Serves: **6-8**

■ Preparation and Cooking Time: **3 hours**

材料：
光雞1隻重約2½斤（1.5千克）
葱1棵，切葱花
紅椒½隻，切絲

味料：
炸蒜茸1湯匙
麻醬1湯匙
麻油2湯匙
精鹽1湯匙
砂糖½湯匙
雞粉½湯匙
魚露1湯匙
玫瑰露酒1湯匙
胡椒粉⅒湯匙
薑6片，切絲
葱2條，切粒

Ingredients:
1 chicken (about 1.5kg)
1 stk scallion, roughly chopped
½ red chili pepper, shredded

Seasonings :
1 tbsp garlic, minced and deep fried
1 tbsp sesame paste
2 tbsp sesame oil
1 tbsp salt
½ tbsp granulated sugar
½ tbsp chicken powder
1 tbsp fish sauce
1 tbsp rose wine
⅒ tbsp pepper
6 slices ginger, shredded
2 stks scallion, chopped

TIPS 偉師傅的專業指導

洗光雞時，要將雞尾肛門的裏外徹底清潔，否則雞熟後會有一股異味。

Make sure to wash the chicken, especially the ass thoroughly. Otherwise, the chicken will have an unpleasant smell after cooking.

做法：
1. 光雞挖去油、肺、喉管，洗淨，瀝乾。
2. 拌勻味料，均勻地抹遍雞的內外，再將全部味料放入雞裏，醃2小時。
3. 將光雞放入窩形碟裏，雞胸向上，置鍋中以大火蒸40分鐘。
4. 取出蒸雞，並把雞裏的汁料倒出。濾去汁料的油脂，然後把中火煮汁料，直至雞汁濃縮成約1碗（約6兩），澆在雞件上，撒上葱花和紅椒絲便成。

Method:
1. Trim all fat from the chicken. Remove the lung and throat. Wash and drain.
2. Blast the chicken with seasonings thoroughly. Stuff the chicken with seasonings and marinate for 2 hours.
3. Put the chicken in a deep plate with the chicken breast facing up. Steam for 40 minutes in high heat.
4. Take out the chicken. Drain the juice and keep it. Drain the fat from the juice. Boil the juice and reduce it to about 225g under medium fire. Chop the chicken. Pour the juice over the chicken. Garnish with scallion and red chili pepper.

巧製奇津雞

Cajun **Chicken**

份量：**6-8 位用** ■製作時間：**3小時**
■ Serves: **6-8**
■ Preparation and Cooking Time: **3 hours**

材料：

光雞1隻重約2½斤（1.5千克）
奇津粉1湯匙
葱絲1條
芫荽葉1棵

味料A：

紅麴米1湯匙，丁香3粒，八角
4粒，香葉10片，沙薑6粒，草
果2粒，果皮¼棵，甘草4片，
桂皮1小塊，沙仁2粒

味料B：

芫荽2棵，生葱3條
生薑4兩（150克），香茅2條
乾葱4粒，洋葱½個
蒜肉4粒
湯骨1斤（600克，汆水）
清水4斤（2.4千克）
生抽王2斤（1.2千克）

味料C：

冰糖1¼斤（750克）
精鹽4兩（150克）
玫瑰露酒1湯匙
紹酒1湯匙

Ingredients:

1 chicken (about 1.5kg)
1 tbsp Cajun seasoning
1 stk scallion, shredded
1 stk parsley

Seasonings A:

1 tbsp red yeast rice, 3 cloves,
4 star anise, 10 bay leaves, 6
zedoary, 2 brown cardamom
pods, ¼ dried tangerine peel, 4
liquorices, 1 pcs cinnamon, 2 pcs
cardamom

Seasonings B:

2 stks parsley, 3 stks scallion
150g ginger, 2 lemon grass
4 cloves shallot, ½ onion
4 cloves garlic
600g pork bone, blanched
2.4kg water
1.2kg light soy sauce

Seasonings C:

750g rock sugar
150g salt
1 tbsp rose wine
1 tbsp Shaoxing wine

↘ TIPS 偉師傅的專業指導

油雞斬件時，先將油雞分
割，抹上奇津粉，再切塊，
讓雞更好地吸收奇津粉。

Chop the chicken before
adding the Cajun seasoning.
After absorbing the Cajun
seasoning, the meat will
become more tasteful.

做法：

1. 將味料A放進布袋裏，備用。
2. 燒熱鍋，下油爆香味料B，加入其餘材料，放入味料A，煮滾，改用慢火熬煮2小時，加入味料C，待糖和鹽完全融化，便成油雞水。
3. 把雞挖去油脂、肺、喉管，洗淨，瀝乾，放入已煮滾的油雞水裏，待雞充滿水份後撈起，將水倒回鍋內，再放回鍋內讓水份留入雞腔，如是者重複5次後，熄火，浸25-30分鐘。
4. 取出雞，待放涼後斬件，鋪上奇津粉，撒上葱絲和芫荽葉便成。

Method:

1. Put seasonings A in a cotton cloth bag. Set aside.
2. Heat the wok with oil. Sauté seasonings B. Add seasonings A and bring to boil. Reduce the fire and simmer for 2 hours. Add seasonings C. Cook until all the salt and granulated sugar is dissolved. This is the chicken brine.
3. Trim all chicken fat. Remove the lung and throat. Wash and drain. Bath in the chicken brine. Take out the chicken when it is soaked with liquid. Repeat 5 times. Turn off the fire and leave the chicken in the brine for 25-30 minutes.
4. Take out the chicken. Leave to cool. Chop the chicken. Drizzle with Cajun seasoning. Garnish with scallion and parsley. Serve.

椒鹽雞軟骨

Sautéed Chicken Soft
Bone in **Spicy Salt**

份量：**6-8位用** ■製作時間：**65分鐘**
■ Serves: **6-8**
■ Preparation and Cooking Time: **65 minutes**

材料：
冰鮮雞軟骨1斤（600克）
炸粉4兩（150克）
葱1條．切絲
紅椒1隻，切粒
椒鹽½茶匙，灑面

味料：
蒜茸汁2湯匙
薑汁1湯匙
胡椒粉⅒湯匙
玫瑰露酒1湯匙
美極鮮露2湯匙
砂糖½湯匙
雞蛋1隻，打勻

Ingredients:
600g chicken soft bone
150g frying powder
1 stk scallion, shredded
1 red chili pepper, chopped
½ tsp pepper salt, for drizzling

Seasonings :
2 tbsp garlic juice
1 tbsp ginger juice
⅒ tbsp pepper
1 tbsp rose wine
2 tbsp Maggi seasoning
½ tbsp granulated sugar
1 egg, beated

TIPS 偉師傅的專業指導

炸雞軟骨時，油要達至七成熱。可先放1粒雞軟骨測試，若看見雞軟骨四周出現小泡沫就，便可放下其他雞軟骨。但如雞粒四周起大氣泡，則表示油溫燒過熱了。

Make sure the oil is hot enough. Test the oil by putting one piece of bone into the wok first. If there are tiny bubbles coming up, the oil is ready. If there are big bubbles coming up, the oil is too hot.

做法：

1. 雞軟骨解凍，洗淨，瀝乾，放入味料裏醃1小時。
2. 將雞軟骨放入炸粉裏拌勻。
3. 燒熱油，放進雞軟骨，以中火炸硬，表面呈金黃色，瀝油，灑上椒鹽。
4. 把雞軟骨放碟上，加葱絲和紅椒粒伴吃，便成。

Method:

1. Defrost the chicken soft bone. Wash and drain. Marinate with seasonings for an hour.
2. Coat the chicken bones with frying powder.
3. Heat the wok with oil. Deep fry the bones in medium heat. Drain the bones when they turn golden brown. Drizzle with pepper salt.
4. Put the bones on a plate. Garnish with scallion and red chili pepper.

鮑汁鳳爪

Chicken Feet in
Abalone Sauce

份量：**10-12位用** ■製作時間：**2小時10分鐘**
■ Serves: **10-12**
■ Preparation and Cooking Time: **2 hours 10 minutes**

材料：
冰鮮大鳳爪1斤（600克）
花生4兩（150克）
蔥1棵，切絲
紅椒¼隻，切絲

味料A：
凍白滷水2斤（1.2千克）（參閱第123頁）

味料B：
清湯10兩（375克）
蠔油1湯匙
雞粉½湯匙
精鹽¼湯匙
砂糖½湯匙
老抽¼茶匙
麻油¼茶匙

芡汁：
生粉¼湯匙＋清水2湯匙（調勻）

味料C：
生薑2片
蒜肉1粒
八角1粒
洋蔥¼個

Ingredients:
600g chicken feet
150g peanut
1 stk scallion, shredded
¼ red chili pepper, shredded

Seasonings A:
1.2 kg white marinade (refer to p.123)

Seasonings B:
375g broth
1 tbsp oyster sauce
½ tbsp chicken powder
¼ tbsp salt
½ tbsp granulated sugar
¼ tbsp dark soy sauce
¼ tbsp sesame oil

Thickening:
¼ tbsp cornstarch + 2 tbsp water

Seasonings C:
2 pcs ginger, slices
1 clove garlic
1 star anise
¼ onion

TIPS 偉師傅的專業指導
1. 煮花生和鳳爪切勿蓋上鍋蓋，否則花生會退皮；鳳爪的外皮會破裂。
2. 鮑汁並非指鮑魚裏煮出的汁，而是指用來做鮑魚芡汁的醬料，這裏是指味料B和C。如果想要效果更好，不妨用清湯和蠔油。
1. Cook the peanuts and chicken feet without coving a lid. To preserve their skins in good appearance.
2. Abalone sauce here is not the sauce made from abalone. Abalone sauce here means one of an ingredient in the thickening agent for abalone. In this recipe, it refers to seasonings B and C. If you want abalone sauce tastes better, add broth and oyster sauce.

做法：
1. 把花生洗淨後，注入1斤（600克）清水、放進1粒八角和½湯匙精鹽，一同煮滾，改用慢火煮1小時，熄火，在汁裏浸泡30分鐘，便成腍花生。
2. 把鳳爪解凍，放入滾水裏以慢火煮5分鐘，熄火，浸30分鐘，泡冷水，涼凍後用剪刀剪去趾甲，沖清水後放入味料A裏，浸滷10小時。
3. 燒熱鍋，下少量油鍋爆香味料C，加入味料B煮滾，轉慢火煮5分鐘，然後將材料取出，加入鳳爪煮滾，改用慢火煮1分鐘。
4. 隔去花生的水份，鋪在碟上，再將鳳爪連汁放上，加蔥絲和紅椒絲便成。

Method:
1. Wash the peanuts and boil with 600g water. Add 1 star anise and ½ tbsp salt. When the mixture comes to boil, reduce the fire and simmer for an hour. Turn off the fire and leave the peanuts in the water for 30 minutes. The peanuts will become very soft.
2. Defrost the chicken feet. Simmer in boiling water for 5 minutes. Turn off the fire and bath for 30 minutes. Rinse in cold water. Put aside and leave to cold. Cut off its nip and rinse again. Marinate with seasonings A for 10 hours.
3. Heat the wok with oil. Sauté seasonings C. Add seasonings B and bring to boil. Reduce the fire and simmer for 5 minutes. Drain the ingredients. Add the chicken feet and simmer for 1 minute.
4. Drain the peanuts. Put them on a plate. Put the chicken feet and the sauce over the peanuts. Garnish with scallion and red chili pepper.

咖喱脆皮雞
Crisp Curry Chicken

份量：**8-10位用** ▦製作時間：**3小時**

▦ Serves: **8-10**

▦ Preparation and Cooking Time: **3 hours**

材料：

光雞1隻重約2¼-2½斤（1.35-1.5千克）

生油2碗（400克）

味料：

蒜茸2湯匙

薑茸1湯匙

洋葱茸⅙湯匙

咖喱粉1湯匙

五香粉⅒湯匙

花生醬½湯匙

清湯6湯匙

精鹽1½湯匙

砂糖1湯匙

雞粉1湯匙

花奶2湯匙，後下

椰漿2湯匙，後下

上皮料：

麥芽糖1湯匙

浙醋1湯匙

滾水3兩（113克）

紹酒½湯匙

Ingredients:
1 chicken (about 1.35-1.5kg)
400g peanut oil

Seasonings :
2 tbsp garlic, minced
1 tbsp ginger, chopped
⅙ tbsp onion, chopped
1 tbsp curry powder
⅒ tbsp five spices powder
½ tbsp peanut butter
6 tbsp broth
1½ tbsp salt
1 tbsp granulated sugar
1 tbsp chicken powder
2 tbsp evaporate milk, use later
2 tbsp coconut milk, use later

Brush-On Glaze:
1 tbsp maltose
1 tbsp red vinegar
113g hot water
½ tbsp Shaoxing wine

TIPS 偉師傅的專業指導

用焗爐燒出來的雞，色澤未必均勻，如先蓋錫紙烘烤熟後才拉油，可確保雞皮顏色平均。

Color of the roast chick will look much better and even by poaching the chicken with oil after cooking in the oven.

做法：

1. 把上皮料拌勻。

2. 將味料裏的蒜茸、薑茸、洋葱茸爆香，加入咖喱粉和五香粉炒香，再與其他味料一同煮滾。

3. 光雞挖去油脂、肺、喉管，洗淨瀝乾，將味料加入雞腔裏抹勻，用鋼針（鴨尾針）縫口，汆水，取出，均勻抹上上皮料，掛起，風乾2小時。

4. 預熱焗爐，把光雞放在烘烤盆裏的鐵架上，蓋上錫紙，放入焗爐中以小火約140℃烘烤25-30分鐘。

5. 取出燒雞，稍涼後拉油，炸至雞皮呈金黃色、口感甘脆時拔針，倒出雞裏的汁液，隔渣，並將原汁放碟底，雞斬件，放上面便成。

Method:

1. Mix the brush-on glaze ingredients.

2. Sauté garlic, ginger and onion to fragrant. Add curry powder and five spices powder, sauté to fragrant. Add the rest of the seasonings and bring to boil.

3. Trim all chicken fat. Remove the lung and throat. Blast the seasonings in the cavity. Sew the chicken with cooking needle. Blanch the chicken. Take out. Blast the chicken with brush-on glaze. Hang up. Air-dry for 2 hours.

4. Preheat the oven. Put the chicken in a roasting pan. Cover the chicken with an aluminum foil. Roast in oven at 140℃ for 25-30 minutes.

5. Take out the chicken. Leave to cool for a while. Poach the chicken with oil until golden brown. Remove the cooking needle. Drain the juice and pour it on a plate. Chop the chicken. Place on the same plate. Serve.

金蒜蠔王燒雞

Roasted Chicken with
Garlic in **Oyster Sauce**

份量：**8-10位用** ■製作時間：**3小時**
■ Serves: **8-10**
■ Preparation and Cooking Time: **3 hours**

材料：
光雞1隻重約2¼-2½斤（1.35-1.5千克）
生油2碗（200克）
炸蒜茸2湯匙

味料：
蒜茸6粒
精鹽1湯匙
砂糖2湯匙
雞粉1湯匙
五香粉⅙湯匙
蠔油2湯匙

上皮料：
麥芽糖1湯匙
浙醋1湯匙
滾水3兩（113克）
紹酒½湯匙

Ingredients:
1 chicken (about1.35-1.5kg)
200g peanut oil
2 tbsp garlic, minced and fried

Seasonings:
6 cloves garlic, minced
1 tbsp salt
2 tbsp granulated sugar
1 tbsp chicken powder
⅙ tbsp five spices powder
2 tbsp oyster sauce

Brush-On Glaze:
1 tbsp maltose
1 tbsp red vinegar
113g hot water
½ tbsp Shaoxing wine

↘ TIPS 偉師傅的專業指導

在雞皮搽上皮料時，要先洗淨雞皮，以防有味料留在雞皮上，影響上皮料的吸收，以至燒雞色澤不均勻。

Wash the chicken skin before blasting with brush-on glaze. This is to make sure no seasonings left on the skin which may affect the taste and color after roasting.

做法：
1. 把上皮料和味料分別拌勻，備用。
2. 光雞挖去油脂、肺和喉管，洗淨，氽水取出，在雞腔裏均勻地抹上味料，洗淨雞皮，再抹上上皮料，倒掛，風乾2小時。
3. 焗爐預熱，把光雞放上盆裏的鐵架，雞胸向上，蓋上錫紙，放入焗爐中以140℃烘烤25-30分鐘。
4. 取出燒雞，稍涼後斬件上碟，灑上炸蒜茸便成。

Method:
1. Mix the brush-on glaze ingredients.
2. Trim all chicken fat. Remove the lung and throat. Blast the seasonings in the cavity. Sew the chicken with cooking needle. Blanch the chicken. Take out. Blast the chicken with brush-on glaze. Hang up. Air-dry for 2 hours.
3. Preheat the oven. Put the chicken in a roasting pan with the breast facing up. Cover the chicken with an aluminum foil. Roast in oven at 140°C for 25-30 minutes.
4. Take out. Leave to cool for a while. Chop the chicken. Drizzle with fried shallot.

香麻酥雞件

Spicy Chicken **Nuggets**

份量：**6-8位用** 製作時間：**70分鐘**
Serves: **6-8**
Preparation and Cooking Time: **70 minutes**

材料：
冰鮮雞腿肉1斤（600克）
芫茜1棵
紅椒絲½隻
泰式雞醬3湯匙

味料：
乾葱茸3粒
蒜茸6粒
薑汁1湯匙
五香粉¼茶匙
胡椒粉¼茶匙
紹酒1湯匙
玫瑰露酒1湯匙
美極鮮露4湯匙
砂糖1湯匙

蘸炸料：
蛋清1隻，打勻
粗粒地瓜粉4兩（150克）
七味粉1茶匙
生白芝麻2湯匙

Ingredients:
600g chicken leg
1 stk parsley
½ red chili pepper, shredded
3 tbsp Thai sweet chili sauce

Seasonings:
3 cloves shallot, minced
6 cloves garlic, minced
1 tbsp ginger juice
¼ tbsp five spices powder
¼ tsp pepper
1 tbsp Shaoxing wine
1 tbsp rose wine
4 tbsp Maggi seasoning
1 tbsp granulated sugar

Batter:
1 egg, use the egg white, beated
150g sweet potato starch
1 tbsp Shichimi
2 tbsp white sesame

◤ TIPS 偉師傅的專業指導

雞腿肉抹上蘸炸料5分鐘後才炸，可使粉材料緊貼雞肉，炸時不易脫落。

To avoid the batter falls off, coated chicken leg needs to rest for 5 minutes before deep frying.

做法：

1. 將雞腿解凍，清洗乾淨，瀝乾，在上面割井字紋，要將筋割斷，但小心不要把雞皮割破，均勻地抹上味料，醃60分鐘。

2. 把蛋清打勻，加入雞腿肉，拌勻，抹上其餘蘸炸料，在碟上放5分鐘，待雞腿肉由黃色變成淺啡色。

3. 把油燒至7成熱，一隻一隻地放入雞腿，用中火炸硬，以至表面呈金黃色，取出。改用大火，放回雞塊，用油炸約半分鐘，取出。

4. 將雞皮向上，切件，放碟裏，放上芫茜和紅椒絲，便可蘸泰式雞醬吃。

Method:

1. Defrost the chicken legs. Wash and Drain. Slightly make some crosses on the skin. Marinate with seasonings for 60 minutes.

2. Whisk egg white. Brush chicken legs with egg white. Coat with the remain ingredients of batter. Leave for 5 minutes until the color of chicken legs changes from yellow to light brown.

3. Heat oil to medium hot. Fry chicken legs one by one. Remove from oil when the skin turn into golden brown. Fry the chicken legs again using high heat for 30 seconds.

4. Chop the chicken. Arrange the chicken legs on a plate. Garnish with scallion and red chili pepper. Serve with Thai sweet chili sauce.

紅燒乳鴿

Roasted **Pigeon**

份量：**4位用** ■製作時間：**8.5小時**

■ Serves: **4**

■ Preparation and Cooking Time: **8.5 hours**

材料：
冰鮮大鴿2隻重約1¼斤（750克）

浸汁：
凍白滷水2斤（1.2千克）（參閱第123頁）

味料：
燒爐乳豬上皮料1份（參閱第122頁）

蘸食料：
熟准鹽½茶匙
噲汁1湯匙

Ingredients:
2 pigeons (about 750g)

Marinade :
1.2kg white marinade (refer to p.123)

Seasonings:
Brine for roasting suckling pig (refer to p122)

Dipping:
½ tbsp pepper salt, sautéed
1 tbsp Worcestershire sauce

做法：

1. 把大鴿解凍，挖去油脂、肺、喉管，洗淨，放入大滾水裏浸20分鐘，取出泡冷水，再放到味料裏泡1小時，取出，泡熱水，在鴿皮上均勻地抹上皮料，風乾2小時。

2. 把2碗油（約400克）燒至180℃，放入乳鴿，不斷攪動，直至鴿皮變成金黃色。

3. 取出乳鴿，斬件，以准鹽和噲汁蘸食。

Method:

1. Defrost the pigeon. Trim the fat. Remove the lung and throat. Blanch for 20 minutes. Marinate with seasonings for 6 hours. Rinse with hot water. Blast with brush-on glaze. Air -dry for 2 hours.

2. Boil 400g of oil to 180℃. Add the pigeon and stir continuously until golden br own.

3. Take out the pigeon. Chop into pieces. Serve with pepper salt and Worcestershire sauce.

TIPS 偉師傅的專業指導

1. 也可把乳鴿放進浸泡貴妃雞的浸汁裏，約泡半小時，否則味道會太濃太鹹。

2. 炸乳鴿時要不停攪動，是為了防止乳鴿黏着鍋底，以至炸焦。

3. 熟准鹽可在超市買現貨，如自製可用幼鹽2湯匙，五香粉⅒茶匙拌勻，用白鑊慢火炒香，待凍後用玻璃樽或膠盒存放便成。

1. Pigeon can also be marinated in the brine for marinating mistress chicken. It should only marinated for an hour. Otherwise, the meat will become too salty.

2. Keep stirring the pigeon during deep-frying. This is to prevent over burnt as it may stick to the wok.

3. Five spices salt can be found in supermarket. Do it yourself you like. Mix 2 tbsp salt and ⅒ tsp five spices powder in a bowl well. Then panfry them in a wok until fragrant. Take out and let cool. Store in a glass bottle or plastic box.

蒜香乳鴿煲

Braised Pigeon with **Garlic**

份量：**2-4位用** ■製作時間：**2小時** ■工具：**7吋瓦煲1個**

■ Serves: **2-4**

■ Preparation and Cooking Time: **2 hours**

■ Utensil: **1 clay-pot, 7-inch in diameter**

材料：
BB鴿2隻共12兩（450克）
蔥絲1棵
紅椒¼隻，切絲

香料：
蒜片5粒
蔥2棵，切段
薑5片
洋蔥½個
生油2湯匙

味料：
油雞水1份（參閱第124頁）

Ingredients:
2 baby pigeons (about 450g)
1 stk scallion
¼ red chili pepper, shredded

Herbs and Spices:
5 slices garlic
2 stks scallion, sectioned
5 slices ginger
½ onion
2 tbsp peanut oil

Seasonings:
Chicken brine (refer to p.124)

↘ TIPS 偉師傅的專業指導

1. BB鴿放在油雞水裏浸熟，已成豉油王乳鴿。

2. BB鴿是指才出生5-6天的鴿子，乳鴿則是出生再久一些的鴿子。現在香港的菜市場售賣活家禽的檔口很難找到BB鴿了，如果沒法弄到BB鴿，不妨以乳鴿代替。

1. As long as the baby pigeon finished cooking in the chicken brine, it is already soy sauce pigeon.

2. Baby pigeon refers to pigeon which is 5-6 days after birth. Pigeon is bit older than baby pigeon. Pigeons can be a substitution when baby pigeon is not available.

做法：

1. 洗淨BB鴿，放入已煮滾的油雞水裏，待鴿身充滿水後撈起，再放回油雞水裏，再重複一次，熄火，泡在湯汁裏浸15分鐘，取出，稍涼後斬件，切成四塊。

2. 把香料放在瓦煲裏爆香，加入BB鴿，排好，再將2湯匙油雞水拌勻，淋在BB鴿上，加入蔥絲，紅椒絲，蓋上煲蓋，用大火煮1分鐘便成。

Method:

1. Wash the pigeons. Poach in the boiled chicken brine until the pigeons are fully soaked with liquid. Take them out. Drain. Repeat the process. Turn off the fire. Bath the pigeons in the brine for 15 minutes. Leave to cool for a while. Chop into 4 pieces.

2. Sauté the herbs and spices in a clay-pot to fragrant. Arrange the pigeons in the clay-pot. Pour 2 tbsp of chicken brine over the pigeon. Cook for 1 minute under big fire.

泰汁鮮果墨魚仔

Baby Squid in Thai Sauce with **Fresh Fruit**

份量：**6-8位用** ■製作時間：**10小時5分鐘**

■ Serves: **6-8**

■ Preparation and Cooking Time: **10 hours 5 minutes**

材料：
冰鮮墨魚仔1磅（454克）
生啤梨2個

味料：
凍白滷水2斤（1.2千克）（參閱第123頁）

泰汁：
泰式雞醬6湯匙
魚露2湯匙
太白司高辣椒汁¼湯匙
麻油2湯匙
甘筍¼個（切絲）
洋葱½個（切絲）
指天椒5隻（切絲）
乾葱肉5粒（切絲）
香茅2條（切絲）
鮮薄荷5片（切絲）
泰國青檸汁2個

Ingredients:
454g baby squid
2 pears

Seasonings:
1.2kg white marinade (refer to p.123)

Thai Sauce:
6 tbsp Thai sweet chili sauce
2 tbsp fish sauce
¼ tbsp Tabasco
2 tbsp sesame oil
¼ carrot, shredded
½ onion, shredded
5 red cluster peppers, shredded
5 cloves shallots, shredded
2 lemongrass, shredded
5 fresh mint leaves, shredded
2 Kaffir limes from Thai, juiced

◺ TIPS 偉師傅的專業指導

1. 泰汁含酸性，故不能用銻和鋁質器皿盛載，否則會引至食物氧化，產生有害物質。

2. 鮮果可選多款配搭，如富士蘋果、哈蜜瓜、奇異果、菠蘿、士多啤梨、石榴、車厘子等等。選取硬身爽口為主。

1. Do not serve the salad in antimony or aluminum bowl. The Thai sauce is acidic and it can oxidize with metal to release poisonous materials.

2. Any kinds of fruit can be suited for this recipe. For example, Fuji apple, honey melon, kiwi, pineapple, strawberry, cherry etc. I advise to use the firm fruit is better than using a soft fruit.

做法：

1. 把啤梨去皮、去核，切成骰子般大小，泡在淡鹽水裏，備用。

2. 把小墨魚解凍、洗淨，放入滾水裏攪勻，待沸騰，便取出泡冷水，再放入味料裏浸10小時。

3. 將墨魚仔、啤梨粒取出，瀝乾，加入泰汁拌勻便成。

Method:

1. Peel the pears. Seed. Cut into dices. Soak in salted water. Set aside.

2. Defrost the squid. Wash. Blanch. Marinate with seasonings for 10 hours.

3. Drain the baby squid and the pears. Toss together with the Thai sauce. Serve.

沙薑墨魚仔

Baby Squid in
Zedoary Sauce

份量：**6-8位用** ■製作時間：**10小時15分鐘**
■ Serves: **6-8**
■ Preparation and Cooking Time: **10 hours 15 minutes**

材料：
冰鮮墨魚仔1磅（454克）
葱1棵，切絲
紅椒¼隻，切絲

味料A：
凍白滷水2斤（1.2千克）（參閱第123頁）

味料B：
生薑2片
八角1粒
蒜肉1粒
洋葱¼個

味料C：
清湯10兩（375克）
精鹽½湯匙
雞粉½湯匙
麻油½湯匙
沙薑粉½茶匙

Ingredients:
454g baby squid
1 stk scallion, shredded
¼ red chili pepper, shredded

Seasonings A:
1.2 kg white marinade (refer to p.123)

Seasonings B:
2 slices ginger
1 star anise
1 clove garlic
¼ onion

Seasonings C:
375g broth
½ tbsp salt
½ tbsp chicken powder
½ tbsp sesame oil
½ tbsp zedoary powder

◥ TIPS 偉師傅的專業指導

味料B加C熬成後就是沙薑汁，可加倍製作，用來浸泡沙薑八爪魚、豬仔腳、鳳爪等等皆可。

To make zedoary sauce, mix seasonings B and C and bring to boil. This sauce can be used in making octopus in zedoary sauce, pig trotters and chicken feet etc.

做法：

1. 把小墨魚解凍，洗淨，汆水，取出泡冷水，再放入味料A裏浸，滷10小時後瀝乾。

2. 用少量油爆香味料B，加入味料C煮滾，改用慢火煮5分鐘，取出香料。

3. 倒進小墨魚，熄火，浸滷10分鐘，加入葱絲和紅椒絲便成。

Method:

1. Defrost the baby squid. Wash. Blanch. Marinate with seasonings A for 10 hours. Drain.

2. Sauté seasonings B with little oil till fragrant. Add seasonings C and bring to boil. Reduce to low heat and simmer for 5 minutes. Drain and discard the spices and herbs.

3. Add the baby squids. Turn off the fire and marinate for 10 minutes. Garnish with scallion and red chili pepper. Serve.

鮑汁八爪魚

Octopus in
Abalone Sauce

份量：**8-10位用** ■製作時間：**5分鐘**
■ Serves: **8-10**
■ Preparation and Cooking Time: **5 minutes**

材料：
冰鮮八爪魚1磅（454克）約
40-60隻
葱1棵（切絲）
紅椒¼隻（切絲）

味料A：
凍白滷水2斤（1.2千克）（參閱第
123頁）

味料B：
生薑2片
蒜肉1粒
八角1粒
洋葱¼個

味料C：
清湯10兩（375克）
蠔油1湯匙
雞粉½湯匙
精鹽¼湯匙
砂糖½湯匙
老抽¼湯匙
麻油¼湯匙

芡汁：
生粉½湯匙＋清水4湯匙（調勻）

Ingredients:
454g octopus (about 40-60 pcs)
1 stk scallion, shredded
¼ red chili pepper, shredded

Seasonings A:
1.2 kg white marinade (refer to p.123)

Seasonings B:
2 slices ginger
1 clove garlic
1 star anise
¼ onion

Seasonings C:
375g broth
1 tbsp oyster sauce
½ tbsp chicken powder
¼ tbsp salt
½ tbsp granulated sugar
¼ tbsp dark soy sauce
¼ tbsp sesame oil

Thickening:
½ tbsp cornstarch + 4 tbsp water

⊿ TIPS 偉師傅的專業指導

1. 焯八爪魚時，時間要操控得宜，時間太短，可能未熟；時間若太長，肉質又會糜爛。

2. 要待水沸騰時才放八爪魚，拌勻，汆水後要立即取出泡冷水，以保持肉質爽脆。另，每次焯煮的份量不宜太多，約20隻就最好。

1. Be ware of the blanching time. The octopus will be undercooked if the blanching time is short. Or the octopus will be overcooked and affect the meat texture.

2. Remember to blanch the octopus in boiling water. To ensure the meat is crunchy, make sure to bath in cold water as soon as possible. On the other hand, do not blanch more than 20 octopuses in the same time.

做法：

1. 把八爪魚解凍，洗淨，汆水，取出泡冷水，放入味料A裏浸滷10小時。

2. 用少量油爆香味料B，加入味料C煮滾，改用慢火煮5分鐘，取出香料，加入八爪魚，煮滾後熄火泡1分鐘。

3. 取出八爪魚，加葱絲和紅椒絲便成。

Method:

1. Defrost the octopus. Wash. Blanch. Marinate with seasonings A for 10 hours.

2. Sauté seasonings B with little oil till fragrant. Add seasonings C and bring to boil. Reduce to low heat and simmer for 5 minutes. Drain and discard the spices and herbs. Add the octopus.

3. Take out the octopus. Garnish with scallion and red chili pepper. Serve.

鹽香蜜汁燒鱔

Roasted Eel with Savory
Honey Sauce

份量：**8-10位用** ■製作時間：**15分鐘**

■ Serves: **8-10**

■ Preparation and Cooking Time: **15 minutes**

材料：

去骨大白鱔1條重約1½斤（900克）
蜜汁3兩（113克）（參閱第122頁）
食粉 ½ 茶匙
鋼針（鴨尾針）16枝
生油1斤（600克）

味料：

精鹽1湯匙
糖1湯匙
雞粉 ½ 湯匙
五香粉 ¹⁄₁₀ 茶匙
沙薑粉 ⅕ 茶匙
蒜粉 ⅕ 茶匙

上皮料：

燒乳豬上皮料1份（參閱第122頁）

Ingredients:

1 eel (about 900g), boned
113g honey sauce (refer to p.122)
½ tbsp baking soda
16 duck needles
600g peanut oil

Seasonings:

1 tbsp salt
1 tbsp granulated sugar
½ tbsp chicken powder
¹⁄₁₀ tbsp five spices powder
⅕ tbsp zedoary powder
⅕ tbsp garlic powder

Brush-On Glaze:

Brine for Roasting Suckling Pig
(refer to p.122)

◥ TIPS 偉師傅的專業指導

大白鱔可以在風乾後分成4份，用保鮮紙包好，放入冰箱貯存2至3天。

Air-dry the eel and then cut into 4 small pieces. With proper wrapping, we can preserve the eel in good quality for 2-3 days in the refrigerator.

做法：

1. 用1湯匙精鹽把大白鱔的污垢和黏液擦去，然後洗淨，瀝乾，在鱔皮上插針。

2. 在鱔皮上均勻地抹上食粉，翻過來，在鱔肉上均勻地抹味料，並用3枝鋼針橫着穿入鱔頭、中、尾段，掛起風乾1小時，冉把上皮料抹在鱔皮上，風乾2小時。

3. 取出鋼針，剁去頭尾，把鱔身分成4份，再用鋼針把每份鱔肉串成「井」字形，備用。

4. 燒熱鍋，下油，煮至六成熟，放入鱔件，鱔皮向下，以慢火炸熟，取出切塊，抹上蜜汁。

Method:

1. Clean the eel by rubbing the body with 1 tbsp of salt. Wash and drain. Prick the eel with pins.

2. Rub the eel with baking soda. Turn over the eel and blast seasonings on the eel. Insert duck needles horizontally in the head, in the middle and in the tail of the eel. Hang the eel and air-dry for 1 hour. Blast with brush-on glaze and air dry again for 2 hours.

3. Take away the needles. Chop and discard the head and tail of the eel. Cut the body into 4 pieces. Probed each portion with needle again to form a magic square. Set aside.

4. Heat the wok with oil. Fry the eel when the oil is medium hot. Let the skin fry first. Use low heat to complete frying. Blast with honey sauce.

蜜汁燒帶子

Honey Roasted **Scallop**

份量：**10-12位用** ■製作時間：**3小時8分鐘**
■ Serves: **10-12**
■ Preparation and Cooking Time: **3 hours 8 minutes**

材料：

冰鮮帶子1磅（454克）約20-40
隻
方包1磅（454克）
炸蒜茸2湯匙
紅椒粒1隻
蜜汁1碗（200克）（參閱第122頁）

味料：

食粉 ½ 茶匙

調味：

精鹽 ⅛ 茶匙
雞粉 ⅛ 茶匙
砂糖 ¼ 茶匙
麻油 ½ 茶匙
粟粉 1 茶匙

Ingredients:

454g scallop (about 20-40 pcs)
454g white bread
2 tbsp garlic, minced and deep fried
1 red chili pepper
200g honey sauce (refer to p.122)

Seasonings:

½ tbsp baking soda

Flavorings:

⅛ tbsp salt
⅛ tbsp chicken powder
¼ tbsp granulated sugar
½ tbsp sesame oil
1 tbsp cornstarch

◥ TIPS 偉師傅的專業指導

冰鮮帶子的貨源包括南澳洲、西澳洲、加拿大、越南和紐西蘭等，其中尤以南澳洲的品質最佳。

Frozen scallops are imported from South Australia, Western Australia, Canada, Vietnam and New Zealand etc. Scallop from South Australia is the best.

做法：

1. 將每片方包切成五個直徑約1吋的圓形或心形，放入焗爐以慢火100℃烘至乾、脆，備用。

2. 把冰鮮帶子解凍，放入食粉拌勻，醃約30分鐘，取出，洗淨，放入調味料醃3小時。

3. 把帶子放入沸水裏煮滾，離火，浸泡1分鐘，取出，瀝乾。

4. 燒熱鍋，把2碗生油煮至6成熱，加入帶子，炸熟，取出，瀝乾，均勻地抹上蜜汁，放在方包上，再用小火鎗烤帶子面層，燒至有少量焦邊，取出，撒上蒜茸和紅椒粒便成。

Method:

1. Cut the bread into 5 circles. Each circle is 1 inch in diameter or heart shape. Toast in oven at 100°C over low heat till crispy.

2. Defrost the scallops. Marinate with baking soda for 30 minutes. Wash. Marinate with flavorings for 3 hours.

3. Boil the scallops in hot water. Remove from heat and marinate for 1 minute. Drain and dry.

4. Heat the wok with 2 bowls of peanut oil till medium heat. Deep fry the scallops. Drain and dry. Glaze with honey sauce. Place them on the bread. Lightly burn surface of the scallop with a fire torch. Drizzle with garlic and red chili pepper. Serve.

【涼菜小食】

APPETIZER

葱花芥末海蜇頭

Jelly Fish
with Mustard and Scallion

份量：**20位用** ■製作時間：**4小時**
■ Serves: **20**
■ Preparation and Cooking Time: **4 hours**

材料：

海蜇頭2斤（1.2千克）
白醋4兩（150克）
葱花白4棵（4湯匙）
紅椒1隻，切絲

味料：

砂糖3湯匙
精鹽1湯匙
雞粉1湯匙
麻油2湯匙
豆瓣醬1湯匙
黃芥末1湯匙
熟芝麻1湯匙

Ingredients:

1.2kg salted jelly fish
150g white vinegar
4 stks scallion, white portions
1 red chili pepper, shredded

Seasonings:

3 tbsp granulated sugar
1 tbsp salt
1 tbsp chicken powder
2 tbsp sesame oil
1 tbsp broad bean paste
1 tbsp English mustard
1 tbsp sesame, sautéed

↘ TIPS 偉師傅的專業指導

切海蜇頭時，注意它是否有硬塊、沙子或是像燒焦一樣的裙邊，如有，必須切走。

Discard any rims and sands, if found.

做法：

1. 用清水把海蜇頭的鹽份洗掉，再放入清水裏浸3小時，每隔1小時更換清水1次，瀝乾。

2. 往海蜇頭中澆白醋，拌勻，放進大約4斤（2.4千克）滾水裏拌勻，倒去熱水，泡冷水，然後啤水30分鐘，取出，切成8毫米小方塊，沖洗乾淨，瀝乾。

3. 加入味料和葱白，拌勻，醃15分鐘，瀝乾，加入紅椒絲便成。

Method:

1. Desalt the jelly fish by rinsing with water. Discard the water and soak the jelly fish in fresh water again for 3 hours. Change the water every hour. Drain and dry.

2. Toss the jelly fish with white vinegar. Blanch in 2.4kg hot water. Stir. Drain. Soak in iced water. Rinse under running water slowly for 30 minutes. Cut into 8 mm dices. Rinse. Drain and dry.

3. Marinate with seasonings for 15 minutes. Toss with scallion. Drain. Garnish with red chili pepper. Serve.

琥珀花生

Caramelized Peanuts

份量：**20位用** ■製作時間：**10分鐘**
■ Serves: **20**
■ Preparation and Cooking Time: **10 minutes**

材料：
去皮花生2斤（1.2千克）
白芝麻4湯匙
白醋1湯匙
砂糖8兩（300克）
清水1斤（600克）
生油3斤（1.8千克）

Ingredients:
1.2kg peanut, skinned
4 tbsp white sesame
1 tbsp white vinegar
300g granulated sugar
600g water
1.8kg peanut oil

TIPS 偉師傅的專業指導

1. 做琥珀花生的花生須要去皮，否則，加入糖漿後會黏一起，變成花生糖。

2. 這道小吃又叫芝麻花生。

1. Make sure to peel the peanuts. therwise, the peanuts will stick together and that is peanut candy.

2. This snack is also called "sesame peanut".

做法：

1. 用乾淨的濕毛巾把白芝麻抹乾淨，然後放進白鑊裏用慢火炒香，也可以焗爐烘焙，放涼備用。

2. 在3斤（1.8千克）滾水裏加入白醋和花生，用慢火煮3分鐘，取出，泡冷水，再煮1次，泡冷水，瀝乾。

3. 另備一鍋，把砂糖和清水慢火煮成糖漿，切勿太稠。

4. 燒熱鍋，把生油煮至5成熱，放入花生，輕輕拌勻，改用慢火炸至油溫達8成熱，聽到花生發出吱吱聲，嗅到草腥味，便取出花生，瀝乾，倒入糖漿裏快速拌勻，取出，再倒進已放芝麻的盆子，用筷子拌勻，並將花生粒分隔，放涼便成。

Method:

1. Wipe clean the sesame with a wet towel. Sauté the sesame in a wok over low heat or roast in an oven till fragrant. Leave to cool. Set aside.

2. Blanch the peanuts in a saucepan with 1.8kg water and white vinegar over low heat for 3 minutes. Repeat the process. Drain.

3. Use any saucepan to make syrup. Boil water and granulated sugar over low heat. Do not make the syrup too thick.

4. Heat the wok with oil. Stir the peanuts gently with oil at medium heat. Change to deep-fry the peanuts. Take out the peanuts when they sizzle and smell glassy. Drain and dry. Toss with syrup. Take out and toss with sesame in another bowl. Separate the peanuts and leave to cool. Serve.

酸辣藕片

Lotus Root in Sour
and Spicy Sauce

份量：**20位用** ■製作時間：**40分鐘**
■ Serves: **20**
■ Preparation and Cooking Time: **40 minutes**

材料：
蓮藕2斤（1.2千克）

味料：
蒜茸8粒
陳醋1湯匙
白醋2湯匙
雞粉1湯匙
砂糖3湯匙
生抽2湯匙
麻油2湯匙
花椒油1湯匙

Ingredients:
1.2kg lotus root

Seasonings:
8 cloves garlic
1 tbsp red vinegar
2 tbsp white vinegar
1 tbsp chicken powder
3 tbsp granulated sugar
2 tbsp light soy sauce
2 tbsp sesame oil
1 tbsp Chinese prickly ash oil

↘ TIPS 偉師傅的專業指導

1. 宜選外形長、直和細的蓮藕，藕肉要潔白，沒有裂口或裂紋，否則恐怕暗藏泥砂，清洗較困難。

2. 日本出產的蓮藕品質最佳，當然價錢比中國蓮藕貴3倍。

1. It is better to use slim and small lotus root. Lotus root should look fresh in color and avoid buying those with cracking because soils may be hidden inside the cracking. This makes the cleaning process very difficult.

2. Lotus root from Japan is the best. The price is 3 times more than the lotus root from China.

做法：

1. 蓮藕去皮洗淨、切薄片，放入80℃熱水裏拌勻，泡冷水，涼凍後泡30分鐘冰水，瀝乾。

2. 燒熱鍋，用少量油爆香蒜茸，加入其他味料煮滾，取出，放涼。

3. 放入藕片拌勻，醃10分鐘便成。

Method:

1. Peel the lotus root. Wash and slice thinly. Blanch in 80℃ hot water. Leave to cool. Further bath in ice water for 30 minutes. Drain. Set aside.

2. Heat the wok with oil and sauté garlic to fragrant. Add other seasonings and bring to boil. Rest in a bowl and leave to cool.

3. Toss with lotus root and marinate for 10 minutes.

麻醬青瓜條

Cucumber Strips with Sesame Paste

份量：**20位用** ■製作時間：**15分鐘**

■ Serves: **20**

■ Preparation and Cooking Time: **15 minutes**

材料：
溫室青瓜2斤（1.2千克）
熟芝麻1湯匙

味料：
蒜茸6粒
精鹽½湯匙
砂糖2湯匙
雞粉1湯匙
芝麻醬3湯匙
麻油2湯匙
冰花梅醬1湯匙
辣油½湯匙

Ingredients:
1.2kg green house cucumber
1 tbsp sesame, sautéed

Seasonings:
6 cloves garlic, minced
½ tbsp salt
2 tbsp granulated sugar
1 tbsp chicken powder
3 tbsp sesame sauce
2 tbsp sesame oil
1 tbsp plum sauce
½ tbsp chili oil

TIPS 偉師傅的專業指導

溫室青瓜洗切後，先浸冰水10分鐘，瀝乾，入口更爽脆。

To enhance crunchiness, bath the cucumbers in ice water after cutting. Drain and dry before serve.

做法：
1. 先將溫室青瓜切去首末兩端，剖開成4條長條，挖去瓜瓢。
2. 再把青瓜修切約2吋長條形，沖洗淨，瀝乾。
3. 放入調味料和熟芝麻拌勻，醃10分鐘便成。

Method:
1. Cut off ends of the cucumber. Discard the ends. Divide the cucumber into 4 long sticks. Remove the pulp.
2. Cut the cucumber into 2 inches long. Wash and drain.
3. Toss with sesame. Marinate with seasonings for 10 minutes. Serve.

七味涼瓜

Bitter Melon in Spicy Sauce

份量：**20位用** ■製作時間：**45分鐘**

■ Serves: **20**

■ Preparation and Cooking Time: **45 minutes**

材料：
泰國長形涼瓜2斤（1.2千克）
鹼水 ⅙ 茶匙

味料：
日本七味粉1湯匙
炸蒜肉1湯匙
熟芝麻1湯匙
麻油1湯匙

Ingredients:
1.2kg bitter melon from Thailand
⅙ tsp lye water
Seasonings:
1 tbsp Shichimi
1 tbsp garlic, deep fried
1 tbsp sesame, sautéed
1 tbsp sesame oil

做法：

1. 把涼瓜直剖開成兩半，再用湯匙挖去瓜瓤，每半邊涼瓜再剖開兩半，斜切薄片。

2. 把5斤（3千克）水燒滾，加入鹼水、⅛茶匙，熄火，放入涼瓜片，快速攪匀，取出泡冷水，涼凍後放入冰水裏浸泡30分鐘，取出，瀝乾。

3. 放入味料，拌匀，醃10分鐘便成。

Method:

1. Cut the bitter melon into 2 half. Remove the pulp by a spoon. Cut each slice into half again. Cut each small slice diagonally into thin slices.

2. Boil 3 kg water. Add ⅙ tsp lye water. Turn off the fire. Add the bitter melon slices. Stir Take out and bath in cold water. Water bath in ice water for 30 minutes. Drain and set aside.

3. Marinate with seasonings for 10 minutes. Serve.

TIPS 偉師傅的專業指導

香港常見涼瓜約分兩種：

1. 身粗而短，肉厚起粒，色澤較深，俗稱「雷公鑿」或「大釘」，適用做熱炒或榨汁。

2. 身直形長，肉質較薄，色澤較青，俗稱「長苦」，適宜煎釀或涼拌，屬泰國種，全年都有生產。

There are two types of bitter melon commonly found in Hong Kong.

1. The first type is oblong and bluntly. It has a warty surface and the flesh is thick. The other name is "Lei kung cho" or "Da Ding". This kind of melon is suitable to sauté or juiced.

2. The other one has a narrower shape with pointed ends. It has relatively thinner layer of flesh and greener in color. It is also named as "Cheung Hu". It is popular in Thai and grows all year round.

香麻莴笋絲

Celtuce Shreds in Sesame Sauce

份量：**20位用** ■製作時間：**20分鐘**
■ Serves: **20**
■ Preparation and Cooking Time: **20 minutes**

材料 :

萬筍3斤（1.8千克）

紅椒絲1隻

熟芝麻2湯匙

甘筍½個

味料 :

砂糖2湯匙

精鹽½湯匙

雞粉1湯匙

麻油2湯匙

豆瓣醬1湯匙

泰式雞醬3湯匙

Ingredients:

1.8kg celtuce

1 red chili pepper

2 tbsp sesame, sautéed

½ carrot

Seasonings :

2 tbsp granulated sugar

½ tbsp salt

1 tbsp chicken powder

2 tbsp sesame oil

1 tbsp broad bean paste

3 tbsp Thai sweet chili sauce

◥ TIPS 偉師傅的專業指導

1. 甘筍絲要比萬筍纖細，否則就會太耀眼，掩蓋了萬筍絲的翠綠。

2. 萬筍汆水，必定要迅速，否則會過熟，吃起來不爽脆。

1. The carrot should shred thinner and finer than celtuces. Otherwise, the red carrot will hide the green bamboo sboots.

2. Overcooked celtuces will lose the crispiness. Therefore, do not blanch them too long.

做法 :

1. 萬筍摘葉、去皮、沖洗乾淨，切絲。

2. 甘筍去皮、切幼絲，然後與萬筍絲一同汆水，快速攪勻，取出，泡冷水，涼凍後瀝乾。

3. 加入調味料拌勻醃10分鐘，瀝乾，加入熟芝麻拌勻，面放紅椒絲便成。

Method:

1. Discard the leaves of the celtuces. Peel, wash, and shred.

2. Peel and shred the carrot. Blanch together with the celtuce shreds. Leave to cool. Drain. Set aside.

3. Marinate with seasonings for 10 minutes. Drain and dry. Toss with sesame. Garnish with red chili pepper. Serve.

欖菜四季豆

String Beans
with Preserved Cabbage

份量：**20位用** ▇製作時間：**15分鐘**
▇ Serves: **20**
▇ Preparation and Cooking Time: **15 minutes**

材料：
四季豆2斤（1.2千克）
樽裝欖菜1湯匙
炸菜2兩（75克）

味料：
蒜茸1湯匙
豆瓣醬1湯匙
蠔油½湯匙
砂糖1湯匙
雞粉½湯匙
麻油½湯匙
清湯8湯匙
胡椒粉1/10湯匙
紹酒¼湯匙

Ingredients:
1.2kg green bean
1 tbsp pickled vegetables
75g preserved vegetables

Seasonings :
1 tbsp garlic, minced
1 tbsp soy bean paste
½ tbsp oyster sauce
1 tbsp granulated sugar
½ tbsp chicken powder
½ tbsp sesame oil
8 tbsp broth
1/10 tbsp pepper
¼ tbsp Shaoxing wine

↘ TIPS 偉師傅的專業指導

1. 以前的廚師會將四季豆切好後，汆水至9成熟才炒，這樣便不會有草腥味，色澤較深，吃起來較軟。

2. 炒四季豆時，不能蓋鍋蓋2次，否則豆會變得乾黃。

1. In the past, chefs liked to cut the beans and then blanch them. When almost cooked, they sautéed the beans. This process can remove the unpleasant flavor and better food presentation.

2. When we sauté the beans, we could not cover them twice. Otherwise, the beans will turn yellow.

做法：

1. 四季豆洗淨，切成約1½吋長條。榨菜去根，切碎。

2. 燒熱鍋，爆香蒜茸、豆瓣醬、欖菜和炸菜粒，加入四季豆炒勻，潷酒，快速炒勻，加入其他味料炒勻，蓋鍋蓋，熄火，焗約1分鐘，取走鍋蓋，炒勻。

3. 如四季豆未熟，可加少量清湯，以慢火炒熟便成。

Method:

1. Wash the beans. Cut into 1½ inche long. Remove the bulb of preserved vegetables. Chop.

2. Heat the wok with oil. Sauté the garlic, soy bean paste, pickled vegetables and preserved vegetables till fragrant. Sauté with the beans. Drizzle with some wine and sauté quickly. Add the rest ingredients of the seasonings. Sauté and then covered with a lid. Turn off the fire. Rest for 1 minute. Sauté again.

3. Add little broth to sauté the beans if necessary. Sauté over low heat till cooked.

金磚脆豆腐

Pan Fried **Tofu**

份量：**20位用** ■製作時間：**10分鐘**
■ Serves: **20**
■ Preparation and Cooking Time: **10 minutes**

材料：
百福盒裝滑豆腐2盒
生油3斤（1.8千克）

味料：
麵粉2兩（75克）
粟粉2兩（75克）
粘米粉2兩（75克）
雞粉1湯匙
七味粉1茶匙

蘸料：
淮鹽適量
喼汁適量

Ingredients:
2 packs Tofu
1.8kg peanut oil

Seasonings:
75g flour
75g cornstarch
75g rice flour
1 tbsp chicken powder
1 tbsp Shichimi

Dipping:
Pepper salt to taste
Worcestershire sauce to taste

TIPS 偉師傅的專業指導

炸豆腐最好用新鮮的生油，否則豆腐很快炸黑，色澤也不均勻。

Always remember to use fresh oil in frying tofu. This will give a nice color and appearance. Re-used cooking oil will darken the tofu and gives a nasty appearance.

做法：

1. 拌勻味料，備用。

2. 將滑豆腐倒出，把四面修切齊整，共切16件。

3. 把豆腐放入味料裏小心拌勻，每件豆腐黏滿味料。

4. 燒熱鍋，把生油燒至7成熱，改用慢火，將豆腐一件一件地放入油裏，炸至色澤金黃。

5. 食用時蘸淮鹽和喼汁。

Method:

1. Mix the seasonings well. Set aside.

2. Evenly divide the tofu into 16 pieces.

3. Coat the tofu with seasonings.

4. Heat oil in a wok to medium heat. Turn down the fire to low heat. Fry the tofu one by one until golden brown.

5. Serve with pepper salt and Worcestershire sauce.

葱花木魚凍豆腐

Tofu with Scallion and **Bonito Flakes**

份量：**12位用** ■製作時間：**10分鐘**
■ Serves: **12**
■ Preparation and Cooking Time: **10 minutes**

材料：
百福滑豆腐2盒
日本木魚花1兩（38克）
白葱花2條
熟芝麻1湯匙

淋料：
酸梅醬4兩（150克）（參閱125頁）

Ingredients:
2 packs tofu
38g bonito flakes
2 stks scallion, white section
1 tbsp sesame, sautéed
Seasonings:
150g plum sauce (refer to p.125)

做法：

1. 將滑豆腐倒出，用刀把豆腐四周切齊，每盒豆腐切成6份，分別放入碟中。

2. 把冰梅汁淋在豆腐上。

3. 再放入葱白、熟芝麻和木魚花便成。

Method:

1. Evenly divide each pack of tofu into 6 pieces. Place them in a plate.

2. Pour the plum sauce over the tofu.

3. Garnish with scallion, sesame and bonito flakes.

◥ TIPS 偉師傅的專業指導

1. 木魚花可往日本食品公司購買。

2. 白葱花，即是生葱只要前段白色的部份。

3. 這菜式變化很大，可加入炸菜粒、皮蛋粒、菜脯粒、熟鹹蛋碎、豬肉鬆、花生碎、炸瑤柱、蝦米茸等等，按喜好自由搭配吧！

1. Bonito flakes can buy at Japanese food store.

2. The white section of scallion means the white bottom part of the root.

3. This dish is versatile and can have numerous variations due to personal taste. It can served with preserved vegetables, preserved eggs, pickled vegetables, diced salty eggs, dried shredded pork, peanuts, fried scallop and dried shrimp etc.

冰梅皮蛋酸薑

Preserved Egg with Pickled Ginger in **Plum Sauce**

份量：**2-4位用** ■製作時間：**10分鐘**
■ Serves: **2-4**
■ Preparation and Cooking Time: **10 minutes**

材料：
松花皮蛋2隻
酸薑片12片
炸粉4兩（150克）
雞蛋1隻

蘸食料：
酸梅醬2湯匙（參閱第125頁）

Ingredients:
2 preserved eggs
12 slices pickled ginger
150g frying mix
1 egg
Dipping:
2 tbsp plum sauce (refer to p.125)

做法：

1. 皮蛋剝殼，放入冰箱雪1小時，取出分成六片。

2. 將雞蛋打散，加入炸粉拌勻，加清水調校成粉漿狀，即簡易脆漿。

3. 把皮蛋放在酸薑上，捲成筒狀，用牙籤串好，放入脆漿裏，粘滿脆漿，放入180℃的油鍋裏，炸脆，表面呈金黃色，取出瀝乾，小心取出牙籤。

4. 食用時蘸酸梅醬便成。

Method:

1. Peel the preserved eggs. Put them in the refrigerator for 1 hour. Take out and cut into 6 slices each.

2. Whisk the egg and mix with frying mix. Add water to form batter.

3. Wrap each sliced preserved egg with a pickled ginger slice and fasten with a toothpick. Dip in the batter. Deep fry at 180°C until golden brown. Drain. Remove the toothpick.

4. Serve with plum sauce.

TIPS 偉師傅的專業指導

1. 皮蛋去殼後要冰凍才切，這樣可以防止蛋黃流失，致使不能定型，影響賣相。

2. 如不想把皮蛋放冰箱冷凍，可以隔清水蒸10分鐘再取出，稍凍後已可切片。

1. Cut preserved eggs after chilling. It is because the egg yolk is set and easy to cut. This will give a nice presentation.

2. Instead of chilling the eggs, an alternate way is to steam the preserved eggs for 10 minutes first. Leave to cool and then cut the egg into pieces.

【常用香料和醬料】

SPICES AND SAUCES

常用的香料 Spices

八角 Star Anise

別　名｜ 大茴香

Chinese name: ba jiao, da hui xiang

原產地｜ 產於濕熱帶地區，古時大多進口，現廣東、廣西、雲南、陝西等
地也種植，以陝西較多。

用　途｜ 1. 作佐料時可有開胃作用。
2. 可磨粉，用來醃肉。

Region of Origin: Grow in wet and humid area. It was mostly imported in early years. In China, star anise is grown in Guangdong, Guangxi, Yunnan and Shaanxi etc. Shanxi is the leader in production.

Usage: 1. As condiment and for food appetizing.
2. It can be grounded to powder to marinate meat.

小茴 Cumin

別　名｜ 小茴香

Chinese name: siu wui, siu wui xiang

原產地｜ 產於濕熱帶地區，古時大多進口，現廣東、廣西、雲南、陝西等地也種植，以陝西較多。

用　途｜ 可磨粉，能辟除肉的羶味，進而疏肝開胃。

Region of Origin: Grow in wet and humid area. It was mostly imported in early years. In China, fennel is grown in Guangdong, Guangxi, Yunnan and Shaanxi etc. Shannxi is the leader in production.

Usage: It can be grounded to powder. It helps to remove the odor of the meat. It is very good for liver and stomach.

丁香 Clove

別　名｜ 丁子香

Chinese name: ding xiang, ding zi xiang

原產地｜ 馬來西亞、坦桑尼亞；廣東亦有種植。

用　途｜ 可磨粉，做佐料能增進食慾，食之齒頰留香。

Region of Origin: Grow in Malaysia, Tanzania and Canton.

Usage: It can be grounded to powder. It is a condiment to aid appetizing and food flavoring.

孜然 Caraway

別　名｜ 安息茴香，又稱「阿拉伯茴香」

Chinese name: ziran

原產地｜ 中亞地區、伊朗、新疆、甘肅、河西。

用　途｜ 可磨粉，稱「孜然粉」，可減少肉羶腥味，具暖胃消滯作用。是調製咖喱首要的香料。

Region of origin: Middle Asia area, Iran, Xinjiang, Gansu and Hexi.

Usage: It can be grounded to powder. It can reduce the offensive smell of meat and can warm the stomach and help digestion. It is a major ingredient in making curry.

桂皮 Cinnamon

別　名｜肉桂皮

Chinese name: rou gui, rou gui pi

原產地｜廣東、廣西、雲南等地，以廣西較多。

用　途｜微辛，可磨粉，功用是去油膩、解燥熱及緩和腸胃。

Region of Origin: Grow in Guangdong, Guangxi, Yunnan etc. Guangxi is the leader in production.

Usage: It is bit hot and spicy. It can be grounded to powder. It can remove the fat, coordinate the stomach and make the body feel comfortable.

草果 Cardamom

別　名｜草果仁

Chinese name: cao guo, cao guo ren

原產地｜雲南、廣西、貴州等地。

用　途｜可磨粉，味帶辛辣，能減少肉的腥味，中和燥濕，化解腸胃濕氣。

Region of Origin: In Hunan, Guangxi, Guizhou etc.

Usage: It can be grounded to powder. It is bit hot and spicy. It can reduce the offensive smell in raw meat. It can dispel dampness and coordinate intestines and stomach.

陳皮 Tanderine

別　名｜果皮

Chinese name: chen pi, guo pi

原產地｜湖南、江西、廣東。

用　途｜可磨粉，有助減少肉腥的味，亦可順氣化痰，幫助消化。

Region of Origin: Hunan, Jiangxi, Guangdong.

Usage: It can be grounded to powder. It helps reducing the fishy smell of raw meat, acting as expectorant and allaying cough and help digestion.

甘草 Liquorices

別　名｜生甘草

Chinese name: gan cao, sheng gan cao

原產地｜遼寧

分　類｜甘草條切片後為甘草片，較遜。

用　途｜可磨粉，味甜，可減少肉的羶腥味，亦可增加人體的膽汁，降低膽固醇。

Region of Origin: Liaoning.

Category: Whole liquorices and Sliced liquorices, The sliced one is not as effective as whole liquorices.

Usage: It can be grounded to powder. It can reduce the offensive smell of meat. It can also increase the volume of bile in human body. It helps to treat cholesterol.

沙薑 Zedoary

Chinese name: sha jiang

原產地｜ 廣東、廣西、台灣省。

用　途｜ 可磨粉，能減少肉的羶腥味，亦可驅散脾、胃寒氣，促進消化功能。

Region of Origin: Guangdong, Guangxi and Taiwan

Usage: It can be grounded to powder. It can reduce the offensive smell of meat. It can strengthen the function of spleen, warm the stomach and help the digestive system.

川椒 Sichuan Pepper

Chinese name: chuan jiao

原產地｜ 四川；產於陝西省的稱為「秦椒」，俗稱「花椒」。

用　途｜ 可磨粉，能減少肉腥味，並可防止肉質滋生病菌，還具有暖胃、消滯的作用。

Region of Origin: Sichuan. For pepper that grows in Shanxi province, it is called Qin jiao, the common name is Chinese pepper.

Usage: It can be grounded to powder. It can reduce the offensive smell of meat and prevent the meat from deteriorating by killing the bacteria and prevents their growth. It can also warm the stomach and help digestion.

香葉 Bay leaves

別　名｜ 香艾

Chinese name: xiang ye

原產地｜ 廣東、廣西。

用　途｜ 味淡辛帶甘，可增加肉質鮮甜，亦具有暖胃、消滯、順喉、生津的功效。

Region of origin: Guangdong, Guangxi

Usage: It is mild in taste but it has a distinctive, pungent flavor. It flavors meat and can warm the stomach and help digestion.

胡椒 Peppercorn

別　名｜ 白胡椒粒；壓成粉末，稱「胡椒粉」或簡稱「古月粉」

Chinese name:　hu jiao. It can be grounded to power and the common name is pepper.

原產地｜ 台灣省。

用　途｜ 味帶辛辣，可減少腥羶味，亦能消除胃裏積氣，引發食慾。

Region of origin: Taiwan

Usage: It has a pungent flavor which can reduce the smell of raw meat, expel stomach gas and improve appetite.

薄荷 Mint

別　名｜ 薄荷葉，番花葉

Chinese name: bo he

原產地｜ 江蘇；泰國、越南。

用　途｜ 可減少其他配味料的辛味，激發肉料的鮮味。

Region of origin: Jiangsu, Thailand and Vietnam

Usage: It reduces the pungency of other seasonings and good in bringing out the meat flavor.

香茅 Lemongrass

別　名｜ 檸檬香茅

Chinese name: xiang mao

原產地｜ 廣東、廣西；泰國、越南。

用　途｜ 可磨粉，增加肉質芬芳的香氣，刺激味蕾，增進食慾。

Region of origin: Guangdong, Guangxi, Thailand and Vietnam

Usage: It can be grounded to powder to increase aroma and improve appetite.

南薑 Pine Core Ginger

Chinese name: nan jiang

原產地｜ 廣東；越南、泰國。

用　途｜ 可磨粉，味帶辛辣，減少肉羶腥味，亦能促進消化，引發食慾。

Region of origin:　Guangdong, Vietnam, Thailand

Usage: It can be grounded to powder. It carries a pungent smell. It can reduce the offensive smell of meat, help digestion and improve appetite.

芫荽籽 Coriander Seed

別　名｜ 芫荽銀

Chinese name: yuan sui zi

原產地｜ 廣東、廣西。

用　途｜ 可磨粉，味微辛，可減少肉羶腥味，還具有暖胃，消滯的作用。

Region of origin: Guangdong, Guangxi

Usage: It can be grounded to powder. It has a light pungent smell. It can reduce the offensive smell of meat, help digestion and improve appetite.

黃枝子 Ellis

別　名｜ 黃箕子 Cape Jasmine

Chinese name: huang zhi zi

原產地｜ 中國南部、中南半島、日本、台灣省。

用　途｜ 味帶少苦，色澤偏黃，主要用以調色，也有很多醫療作用，可增進食慾。

Region of origin: South China, Indochina, Japan, Taiwan.

Usage: It has a bitter flavor and yellow in color. It is used for food coloring and has various medical uses. It can improve appetite too.

紅麴米 Red Yeast Rice

別　名 | 紅米

Chinese name: hong qu mi

原產地 | 印度、泰國

用　途 | 主要用於調色。紅米是糙米的一種，外殼大多未被輾磨，口感粗
糙，有豐富的蛋白質、纖維素、礦物質及維他命B。

Region of origin: India, Thailand

Usage: Mainly for food coloring. Red yeast rice is a family member of brown rice which keeps the
outermost layer of the grain. It is chewy and rich in protein, fiber, minerals and vitamin B.

砂仁 Amomum Fruit

別　名 | 縮砂仁

Chinese name: sha ren

原產地 | 馬來西亞吉隆坡，為薑科多年生草本植物的乾燥成熟果實。

用　途 | 微帶辛辣，可減少羶腥味。具有行氣健脾、中和胃氣、消積滯的
作用。

Region of origin: Kuala Lumper, Malaysia It is good in function the spleen, reduce stomach gas and help
digestion.

Usage: It is slightly spicy and hot. It can reduce the offensive smelly of meat.

五香粉 Five Spices Powder

五種藥材磨成粉末，混和而成，香料成份計有桂皮、八角、丁香、沙薑、甘草，如不想麻煩，
可去超市買現成的。

It is a mixed powder with 5 herbs and spices. They are cinnamon, star anise, zedoary, and liquorices.
You can make this powder by yourself or buy instant one from supermarket.

咖喱粉 Curry Powder

咖喱粉以包裝分，有兩大類：(1) 大包斤裝，按每斤算、小包安士裝，按每包算。(2) 有油浸
的玻璃樽裝油咖喱。

以味道和產地細分，則有數百種了，有紅咖喱、黃咖喱、綠咖喱、青咖喱，有產自印度、新加坡、
馬來西亞、海南島、印尼等不同種族的，還有許多餐館的多方配製，真是千變萬化。

在家裏，你可以用原材料咖喱粉，混入其他你喜歡的香料，調配出一種口味獨特，能滿足自己的咖喱。

There are two types of packings: (1) powder in bag: large ones measures by kilograms and small ones
measured by grams. (2) bottled curry paste.

There are hundreds of curry. That can be red curry, yellow curry and green curry. Also there are curries
from different regions, such as India, Singapore, Malaysia, Hoinam Island, Indonesia etc. And there are
homemade curries from different restaurants. You can make your own unique curry flavor by mixing
your favorite herbs and spices with the curry base.

燒烤醬（又名「萬用醬」或「百搭醬」）

BBQ Sauce (All purpose sauce)

煮時切勿分心，鏟醬時盡量不讓它黏底。

Be focus in cooking; stir with a spatula frequently. Don't burn the sauce.

材料A：
生油8湯匙
蒜茸10粒
乾葱茸10粒
陳皮茸1棵

材料B：
海鮮醬1斤（600克）
磨豉醬4兩（150克）
芝麻醬4兩（150克）
蠔油4兩（150克）
南乳1大件
砂糖8兩（300克）
雞粉2兩（75克）
生抽8兩（300克）
清湯8兩（300克）
五香粉1茶匙
麻油1兩（38克）

Ingredients A:
8 tbsp oil
10 cloves garlic, minced
10 cloves shallot, minced
1 tangerine, fine chopped

Ingredients B:
600 g Hoi Sin sauce
150 g soy bean paste
150 g sesame paste
150 g oyster sauce
1 pcs red preserved tofu
300 g sugar
75 g chicken powder
300 g light soy sauce
300 g stock
1 tsp five-spices powder
38g sesame oil

做法：
燒熱油鍋，爆香材料，再和材料B混和煮滾，繼續煮2分鐘，取出，用食物盒盛載，涼凍後放冰箱。

Method:
Heat the wok with oil, sauté ingredients until fragrant. Mix with ingredients B, Bring to boil for 2 minutes. Leave to cool and stored in the fridge.

MATCHING 適用菜式

燒乳豬、叉燒、排骨、燒鵝、鴨、雞、乳鴿等等。
Roast suckling pig, BBQ pork, spare rib, roast goose, duck, chicken, pigeon, etc.

涼菜小食

常用香料和醬料

蜜汁 (熟芽糖)
Honey Sauce

材料
麥芽糖12兩 (450克)
清水6湯匙、冰糖2兩 (75克)
精鹽1茶匙、薑2片

Ingredients:
450g Maltose
6 tbsp water, 75g rock sugar
1 tbsp salt, 2 slices ginger

做法
將清水、冰糖、精鹽和薑片一同放入碗裏，麥芽糖連罐一起放在平底鍋上，隔水蒸溶，直到冰糖全部溶解，把全部材料混合，以慢火煮滾便成。

Method:
Put water, rock sugar, salt and ginger in a bowl. Place the can of maltose in a saucepan. Bain-marine the maltose until dissolved. When the rock sugar is totally dissolved, mix with the maltose. Bring the mixture to boil under small fire.

☒ MATCHING 適用菜式

一般燒製食物，叉燒、排骨、帶子、燒鱔等等。
General roast food such as BBQ pork, spare ribs, scallops and eel etc.

燒乳豬上皮料 (乳豬水)
Brine for Roasting Suckling Pig

材料
白醋4兩 (150克)
浙醋1茶匙
麥芽糖1/10茶匙
玫瑰露酒1/10茶匙

Ingredients:
150g white vinegar
1 tbsp red vinegar
1/10 tbsp maltose
1/10 tbsp rose wine

做法：
將白醋、浙醋、麥芽糖一同放在滾水上，燉至麥芽糖完全溶解，涼凍後加入玫瑰露酒拌勻便成。

Method:
Bring white vinegar, red vinegar and maltose to boil in a pan. When all the maltose has dissolved, set aside. Leave to cool. Add rose wine.

☒ MATCHING 適用菜式

燒、烤、烘烤乳豬，燒乳鴿，燻脆皮雞，燒鱔等等。
It is suitable in roasting suckling pig, roasting pigeon, smoking crispy chicken and roast eel.

燒雞上皮料 (燒雞水)
Brine for Roasting Chicken

材料
麥芽糖1湯匙
浙醋1湯匙
滾水3兩（113克）
紹酒½湯匙

Ingredients:
1 tbsp maltose
1 tbsp red vinegar
113g boiling water
½ tbsp Shaoxing wine

📐 TIPS 偉師傅的專業指導

以上材料如經常使用，可增加份量，製好裝進瓶子裏，存放在冰箱裏，待用時取出，用前先搖一搖。

If the glaze is used frequently, the quantity can be increased. It can be stored in a bottle in the refrigerator. Shake before use.

做法：
將麥芽糖、浙醋和滾水一同攪至麥芽糖溶解，放涼凍，才加入紹酒拌勻便成。

Method:
Add maltose, red vinegar in hot water till all maltose dissolved. Leave to cool. Add the Shaoxing wine.

📐 MATCHING 適用菜式
各款燒雞、脆皮雞、炸子雞、炸乳鴿、燒鰻魚等等。
For different types of roast chicken, crispy chicken, deep fried chicken, roast pigeon and roast eel, etc.

凍白滷水
White Marinade

材料
清水5斤（3千克）、薑片2片
八角2粒、桂皮1小塊
丁香2粒、果皮¼個
草果2粒、沙薑6粒
香葉6片、甘草4片
冰糖2兩（75克）、精鹽4兩（150克）
雞粉1湯匙、玫瑰露酒2湯匙，後下

Ingredients:
3kg water, 2 slices ginger
2 star anise, 1 small pc cinnamon
2 cloves, ¼ tangerine
2 brown cardamom pods
6 pcs zedoary, 6 pcs bay leaves
4 pcs liquorices
75g rock sugar, 150g salt
1 tbsp chicken powder
2 tbsp rose wine, add later

📐 TIPS 偉師傅的專業指導

凍白滷水最好預早熬煮好，放在冰箱裏，用時將渣隔走，才加玫瑰露酒，以免藥材碎末黏在肉料裏。

The white marinade can prepare in advance and store in the fridge. When in use, strain the sauce before adding the rose wine. That will ensure the marinade is clean and nothing will stick on the meat.

做法：
以上材料煮滾，慢火熬煮20分鐘，熄火，放涼凍，加玫瑰露酒拌勻便成。

Method:
Bring all the ingredients to boil. Simmer for 20 minutes. Turn off the fire and leave to cool. Add the rose wine.

📐 MATCHING 適用菜式
燻蹄、鳳爪、豬仔腳及一般冷盤的入味料。
Marinade for smoke knuckle, chicken feet, pork trotters and other cold dishes.

油雞水
Brine for Soy Sauce Chicken

材料 A
紅麴米1湯匙、丁香3粒
八角4粒、香葉10片
沙薑6粒、草果2粒
果皮¼片、甘草4片
桂皮1小塊、沙仁2粒

材料 B
芫荽2棵、生葱3棵
生薑4片、乾葱頭4粒
洋葱½個、蒜肉4粒
湯骨1斤（600克），汆水
清水4斤（2.4千克）
生抽王2斤（1.2千克）

材料 C
冰糖1¼斤（750克）
幼鹽4兩（150克）
玫瑰露1湯匙、紹酒1湯匙

Ingredients A:
1 tbsp red yeast rice, 3 cloves
4 star anise, 10 bay leaves
6 zedogary
2 brown cardamom pods
¼ fried mandarin peel
4 liquorices
1 pcs cinnamon, 2 cardamom

Ingredients B:
2 stk parsley, 3 stk scallion
4 slices ginger, 4 shallots
½ onion, 4 gloves garlic
600g pork bone, scalded
2.4kg water
1.2kg light soy sauce

Ingredients C:
750g rock sugar
150g salt
1 tbsp rose wine
1 tbsp Shaoxing wine

⊿ TIPS 偉師傅的專業指導

因為油雞水是新調校，所以藥材味和滷味就很淡；如想效果好些，可在油雞水熬成後，熄火靜待8小時，才將味料B的湯渣濾走。

As the chicken brine is freshly made, the herb flavor and marinade flavor are light. To have stronger flavor, rest the brine for 8 hours. Then take out the leftovers from the soup.

做法
1. 用少量油爆香材料B，轉入煲裏，加入湯骨、清水、生抽王及材料A（用煲湯隔渣袋裝好），煮滾後以慢火熬2小時。
2. 加入材料C，煮至糖鹽溶解，便成油雞水。
3. 使用前先將材料B的湯渣濾走。

Method:
1. Sauté ingredients B with oil until fragrant. Transfer to a sauce-pan. Add pork bone, water, light soy sauce and ingredients A (packed in a cotton cloth bag). Bring to boil and then simmer for 2 hours.
2. Add ingredients C. Simmer until the salt is dissolved. The chicken brine is ready.
3. Strain the soup before use.

⊿ MATCHING 適用菜式

玫瑰豉油雞、乳鴿、BB鴿、奇津雞等。

For rose soy sauce chicken, soy sauce pigeon, soy sauce baby pigeon and Cajun chicken, etc.

冰花梅醬（酸梅醬）

Plum Paste

材料A
冰糖8兩（300克）、白醋2兩（75克）
清水2兩（75克）

材料B
酸梅4兩（150克，去核）
西檸¼個（去核）

材料C
紅椒粒½隻、酸薑粒1兩，切絲

Ingredients:
300g rock sugar
75g white vinegar
75g water

Ingredients B:
150g plum, seeded
¼ lemon, seeded

Ingredients C:
½ red chili pepper
38g pickled ginger, find disced

◥ TIPS 偉師傅的專業指導

超市有現成酸梅醬出售。
如果想自己親手做，可用2
碗糖醋，加1碗冰花梅醬，
混和煮滾，放涼凍便成。

Ready made plum sauce can
be found in supermarkets.
To make by yourself, bring
2 bowls of sugar vinegar
and 1 bowl of plum paste
to boil. Stir and leave to
cool.

做法

將材料A隔滾水蒸溶，加入材料B，放入攪拌機裏攪至均勻，再轉入鍋裏，以慢火煮滾，放涼凍後加入材料C料，混勻便成。

Method:

Steam ingredients A till dissolved. Add ingredients B. Blend the mixture and stir well. Transfer to a sauce-pan. Simmer till boil. Leave to cool. Add ingredients C. Stir well.

◥ MATCHING 適用菜式

1. 可作鵝、燒鴨、煙鴨胸等等蘸食料。

2. 可作梅子蒸鵝、梅子蒸排骨、酸梅醬等配製料。

1. Dipping for goose, roast duck, smoke duck breast.

2. Ingredients for dishes like steam goose in plum sauce, steam spare ribs in plum paste and plum sauce.

口水辣汁
Hot and Spicy Sauce

材料

砂糖8湯匙、精鹽2湯匙

雞粉2湯匙、陳醋4湯匙

老抽1湯匙、生抽1湯匙

麻油4湯匙、魚露2湯匙

花生醬2湯匙、辣椒油4湯匙

花椒油1湯匙、川椒粉 ⅕湯匙

花椒粉 ⅕湯匙、孜然粉 ⅕湯匙

清水6湯匙

Ingredients:

8 tbsp sugar, 2 tbsp salt

2 tbsp chicken powder

4 tbsp mature vinegar

1 tbsp dark soy sauce

1 tbsp light soy sauce

4 tbsp sesame oil

2 tbsp fish sauce

2 tbsp peanut butter, 4 tbsp chili oil

1 tbsp Chinese prickly ash oil

⅕ tbsp Sichuan chili powder

⅕ tbsp paprika

⅕ tbsp caraway powder

6 tbsp water

◥ TIPS 偉師傅的專業指導

另一製作方法是把全部材料一起調和，然後用慢火煮滾，待涼，當然效果稍遜。

Alternate way is to simmer all ingredients in the same time. The quality, of course, may be less desirable.

做法

1. 可先將砂糖、精鹽、雞粉、川椒粉、花椒粉和孜然粉拌勻。

2. 再加入麻醬、花生醬、麻油、花椒油和辣椒油，一起拌勻。

3. 然後才加陳醋、老抽、生抽、魚露和清水，一起拌勻。

4. 最後用慢火煮滾，待涼。

Method:

1. Mix sugar, salt, chicken powder, Sichuan chili powder, paprika and cumin powder together.

2. Then add peanut butter, sesame paste, sesame oil, Chinese prickly ash oil and chili oil. Stir well.

3. Add mature vinegar, dark soy sauce, light soy sauce, fish sauce and water. Stir well.

4. Simmer to boil. Leave to cool.

◥ MATCHING 適用菜式

口水雞、口水雙脆、夫妻肺片等。

Sichuan spice chicken, Sichuan beef and pork lungs in chili sauce etc.

附錄 | 燒味行話

準備階段

啤 水： 把物料放在水喉下，以慢慢流動的清水漂洗。

捉 水： 泛指把雞浸入味料，讓雞胸腔充水，然後撈起，將水倒回味料內的工序。好處是全雞裏外同的溫度一致，受熱均勻。

拉 油： 把物料掛在滾油上，再用長杓把油澆在物料上，利用油溫把物料炸熟。

上 皮： 廣東廚師的行話，「上」在這裏是動詞，抹上的意思；上皮就是把味料抹在皮上的意思了。

鬆 針： 用針插在豬皮上刺插。

完成階段

打 好： 廣東廚師的行話，將乳豬或將枚肉分割好。

火 雞： 豬頸肉或叉燒旁的燒焦部份。

川辣醬汁
Sichuan Chili Sauce

材料
砂糖4湯匙、精鹽1湯匙
雞粉1湯匙、川椒粉 1/10 湯匙
孜然粉 1/10 湯匙、花椒粉 1/10 湯匙
麻油2湯匙、辣椒油2湯匙
白醋2湯匙、老抽1湯匙
花椒油 1/2 湯匙、清水2湯匙

Ingredients:
4 tbsp sugar, 1 tbsp salt
1 tbsp chicken powder
1/10 tbsp Sichuan chili powder
1/10 tbsp cumin powder
1/10 tbsp paprika
2 tbsp sesame oil
2 tbsp chili oil, 2 tbsp white vinegar
1 tbsp dark soy sauce
1/2 tbsp Chinese prickly ash oil
2 tbsp water

TIPS 偉師傅的專業指導

可以多做幾份，隨時應用。
Prepare a few more for further use.

做法
1. 用攪拌機將材料拌勻。
2. 用慢火煮滾，待涼。
3. 用食物盒盛載，放入冰箱，用時取出。

Method:
1. Blend all ingredients and stir well.
2. Simmer to boil. Leave to cool.
3. Store in food container and put it in the fridge. Use as needed.

MATCHING 適用菜式

麻辣金肚絲、麻辣豬肚絲、麻辣仙掌等等。
Sichuan spicy pork tripe, Sichuan spicy shredded tripe and hot and spicy duck web etc.

Appendix | Jargons

For Preparation

Ber Sui- Rinsing: Rinse item under slow running water.

Dan Sui- Soaking: Generally it is a marinating process of a chicken in brine. The chicken will then drained when the cavity is filled with liquid. The advantage is that the all parts of the chicken receive same temperature.

Lai Yau- Poaching: It is a quick simmering process. Hot oil is poured over the hanging food so that the heat can cook the food.

Sheung Pei- Brushing: Put brush-on glaze all over the food body.

Sung Zum- Pricking: Prick the skin of the pig by a needle to release air bubbles.

After Cooked

Da Ho- Cutting the suckling pig.
Fo Gai- Burnt nibs:The tiny burnt meat balls around the roasted pork.

在家做燒味 Homemade Siu Mei and Luo Mei

編著　Author
陳偉　Chan Wai

編輯　Editor
郭麗眉　Cecilia Kwok

翻譯　Translator
威廉　William

攝影　Photographer
幸浩生　Johnny Han

設計　Designer
任霜兒　劉紅萍　馮麗珍　Annie F　Adrianne Feng　Pancy Liu

出版者　Publisher
萬里機構・飲食天地出版社　Food Paradise Publishing Co., an imprint of Wan Li Book Co Ltd.
香港鰂魚涌英皇道1065號東達中心1305室　Room 1305, Eastern Centre, 1065 King's Road, Quarry Bay, Hong Kong.
電話　Tel: 2564 7511
傳真　Fax: 2565 5539
網址　Web Site: http://www.wanlibk.com

發行者　Distributor
香港聯合書刊物流有限公司　SUP Publishing Logistics (HK) Ltd.
香港新界大埔汀麗路36號中華商務印刷大廈3字樓　3/F, C & C Building, 36 Ting Lai Road, Tai Po, N.T., Hong Kong.
電話　Tel: 2150 2100
傳真　Fax: 2407 3062
電郵　E-mail: info@suplogistics.com.hk

承印者　Printer
凸版印刷（香港）有限公司　Toppan Printing Co (HK) Ltd.

出版日期　Publishing Date
二〇一三年一月第五次印刷　Fifth Printing in January 2013

ISBN　978-962-14-4017-4